Consequences

An Intelligence Officer's War

David Grantham

BLUE EAR BOOKS

Published in 2020 by
Blue Ear Books
7511 Greenwood Ave N, Box 400
Seattle, WA 98103
USA

www.blueearbooks.com

ISBN: 978-0-9844063-1-9

Credits:

Cover Design:
Jan Ballard
Book Composition:
Jennifer Haywood, Blue Ear Books, Seattle, WA
Printer:
Scott Morris, Morris Printing Services, Arkansas City, KS, www.morrisprint.com

Russell,

God Bless!

CONTENTS

PART THREE

PREFACE

This book is a personal account of my short journey through America's longest war. My experience as an Air Force counterintelligence officer in the battle against international terrorism is particular to a time and place, and yet incredibly relevant to ongoing events.

When I arrived in the warzone in the winter of 2006, Iraq was teetering on the razor's edge of all-out civil war. It felt as though a massive black wave was cresting, bearing down on us. The U.S. military desperately needed intelligence to turn the tide. The call for sensitive operators landed me in Kuwait, Afghanistan, and then ultimately at southern Iraq's infamous forwarding operating base, Camp Bucca.

The prison at Camp Bucca is familiar to most as home to the war's most dangerous terrorists and the breeding ground for the Islamic State. It was more. The base was in the Shia-dominated south of Iraq, where our deadly fight with Iran was an open secret. As counterintelligence officers, we were plying perhaps the most dynamic trade in government, and with Al Qaeda in the prison and Iran at the gates, it felt as if we were fighting two wars at once.

This book is a glimpse into the rawness, the pressure, the pleasures and consequences that come with wartime intelligence. The work lent even the most mundane events an exceptional quality, and the common man a unique voice. That is too often overlooked in the world of spy craft. More than just cinematic adventure and international intrigue, more personal than politics or historical analysis,

I hope my story will feel at once familiar to you and altogether different from what you may know. This is the real-life story of an ordinary man turned intelligence officer who found himself in an extraordinary environment.

I haven't written this book from the comforts of retirement or from the ivory tower of academia. That is to say, I am not at an end. The sorrow, the grit, the excitement still surround me, surround us. I write as the winds of time continue to swirl, whipping and pulling my mind and my opinions in different directions. History continues to unfold, and not in the clean and neat way that the phrase suggests.

A context to my story is needed here. By 2006, war planners realized that the sophisticated tactics of previous years that enabled U.S. forces to steamroll the Iraqi military could not overcome poor planning. That prompted the "Surge" in 2007, which brought thousands of U.S. troops back to Iraq to recapture a war thought won. I watched from home in Texas as that surge settled, if for a moment, a troubled environment. A withdrawal that began in 2009 brought a new unraveling. And so, on it went.

The consequences collided with the Arab Spring in 2011 as widespread disenfranchisement and political vacuums fueled all kinds of relationships, both predictable and unorthodox, throughout the war-torn Middle East. Aftereffects of Camp Bucca, which felt distant from current events, flowed forth anew as ISIS shot to prominence. Memories of imprisoned war combatants were once again in the American conscience. Osama Bin Laden was dead, as was Saddam Hussein. Sunni discontent was evident. ISIS had a stable of willing participants spread across Syria and Iraq.

By 2015, ISIS unleashed its bloodletting on Europe, just as the Iran nuclear deal brought with it reminders of the Islamic regime's shadow war in Iraq. Digital propogandists tied to both of America's adversaries fueled extremism across the globe. From San Bernardino to Paris, Brussels to Bali, the consequences of America's war seemed all around us.

The saga continues as of this writing. Abu Bakr al-Baghdadi, a former inmate at Camp Bucca and the first leader of ISIS, was killed in October 2019. Iranian General Qasem Soleimani, the architect of Iran's murderous strategy in Iraq, was killed four months later.

All events recounted in this book occurred to the best of my memory. Some names have been changed and some events adjusted to protect those still in the field. I borrowed details about the operation and airstrike that ultimately killed Abu Musab Al-Zarqawi, leader of Al Qaeda in Iraq, from a July 4, 2016 episode of CNN's "Declassified: Killing Zarqawi." Joby Warrick's *Black Flags: The Rise of ISIS* was helpful in recalling some of the physical features of Camp Bucca.

July 2020
Fort Worth, Texas

PROLOGUE

I rocked back and forth in my ratty brown recliner. It was the same recliner I had purchased ten years earlier, soon after leaving the only house I knew in Tampa, the only city I knew, to join the United States Air Force Office of Special Investigations (OSI). I had accepted my assignment with the military's premier criminal investigative agency in June 2004 and soon after departed for my first duty station, with only a handful of home furnishings and some pocket cash. This cheap recliner had been my first major purchase as a newly commissioned officer. Now, worn and stained from years of abuse and multiple moves cross-country, it sat, much to my wife's consternation, in the living room of our modest North Texas home.

I lumbered out of it now and began lacing up my shoes for a routine afternoon jog. As I gingerly extended my stiff legs into a stretching position, my running partner, Mike, sat unmoved on the sofa. Mike had a near superhuman ability to run for miles with little or no preparation. Before setting out on this day in 2014, we both waited patiently for the Texas sun to descend just a bit more.

The sportscasters on the television had just reissued the same commentary they had offered an hour earlier. The 24-hour news cycle could drive even the most patient person crazy. Even sports, something I rarely tired of, fell victim to the parade of talking heads and incessant, closed-loop conversations.

My infant son played at my feet. A mischievous smile came and went on his face, as he clumsily picked at the laces of my shoes. It amazed me that glimpses of a child's personality could show this early in life. But there it was.

"Food?" I asked, pointing my fingers together toward my mouth in the recognized motion for eating. Immediately he lost interest in my shoes and grunted in agreement. I stood, reached down and scooped him up, took a spot behind the counters that overlooked the living room, and hoisted him into an infant seat. Mike browsed stations on the TV, seemingly lost in thought. Guys don't want to know what's on TV, they just want to know what *else* is on TV, as Jerry Seinfeld said.

Mike paused on some news channel, and we caught the dialogue midstream. Foreign policy analysts and commentators challenged each other's theories on U.S. strategy, terrorism, and the brutal civil war in Syria. Round and round they went. At times the animated discussion seemed genuine, at other points manufactured, as if someone off camera had instructed them to "liven it up." In any case none of them had a solution, partly because they clearly knew little about the facts on the ground. Who was fighting whom. And why.

"What is this, exactly?" Mike said. "Are we involved?"

"No idea. I'm a civilian now," I said, as I fed my son another spoonful of pasta.

The TV conversation turned to the latest development, the formation of a group calling itself the Islamic State of Iraq and Syria, or ISIS. A gaggle of Al-Qaeda rejects or a merger of several terrorist organizations, the broadcaster wasn't entirely sure. But this group was running roughshod over northern Iraq and parts of eastern Syria, publicly executing countless people on their merciless march through the cradle of civilization.

Then I heard a reference to Camp Bucca, and I froze. "Turn that up," I said to Mike.

With my son secured in his chair and now fidgeting with his pasta rather than eating it, I circled back toward the living room. The commentator mentioned Camp Bucca again. Some of the group's top leadership had spent time at the notorious U.S. detention facility, as prisoners of Operation Iraqi Freedom.

Every bad guy spent time there, I mumbled to myself. This wasn't news. But the story grew more precise. They referenced Bucca as the group's breeding ground. Its leader had spent some months in custody there. "Experts called it a jihadist training camp," the reporter now on camera said. "Some call it the birthplace of this group, ISIS."

In an instant, everything came flooding forward. All our work, our experience, our informants, all the intelligence we had collected. The collaboration among Al Qaeda prisoners, the Sunni-Shia divide – all those seemingly disconnected pieces of intelligence – now seemed part of a larger and more dangerous story. Something fell into place in my mind. This seemingly small revelation gave my experience new clarity, even though the violence, the suffering, the mayhem seemed so distant from what I now knew. The news had it partly right. Camp Bucca was less a birthplace than a midwife or nanny. Bucca didn't create ISIS, but it certainly nursed it well in those early years. It gave the organization a healthy start.

In my head echoed the adage that we can choose our decisions, but not the consequences. Eight years after leaving Bucca, I was watching the unintended consequences unravel an already unstable Middle East.

"Well, you ready?" said Mike, interrupting my train of thought.

"Sure," I mumbled, still distracted. My mind continued racing, reconstructing past events, reviewing circumstances and incidents I had once discarded as unnecessary. Perhaps I had missed something. People, colors, and smells resurfaced, as if a door to hidden memories had been flung open. The sensation of Iraq's desert air gently moved over my face; in my mind's eye I could see the prison.

My son's laughter brought me back. I shook my head clear and saw him trying to squirm out of his chair, trying to reach his mother, who had just walked through the door. She smiled and scooped him up.

"Enjoy the run," she said.

PART ONE

CHAPTER 1

Midwife of ISIS

A faded-white thermometer dangled on a tall cement barrier directly outside the window of my dingy office trailer. It was April 2006, and the last few days had reached 100 degrees Fahrenheit. The thermometer was suited for summer, topping out at 125.

The sun sapped one's energy quickly. My clothes held their stench of sweat and wafted it through the air. I teased my linguist that terrorists valued suicide over life not because of religion or U.S. policy, but because of this weather. Abdul was a short, plump Egyptian with a hearty personality. I readied myself, rose from my chair, and reached for my badge. "Here goes," I said. Abdul assumed I was leaving for lunch. "I think I will stay inside with this heat," he said without prompting. "Bring me back a takeout tray, will you?" The comment sounded like a command rather than a request. I thought it curious that voice inflections in the Arab world so often sounded authoritative. I simply nodded, knowing that I wouldn't return to the office for the rest of the day. We were merciless to each other. Smiling to myself, I swung wide the office door and stepped out under the searing midday sun.

I squinted as my eyes adjusted to the sun reflecting off the sand. I scanned the area and eventually located my ride off in the distance. The heavy-duty Gator ATV scheduled to transport me to the prison idled in the makeshift parking area about fifty yards away. I began walking briskly toward it. "You here for Grantham?" I yelled to the young driver as I approached.

"Yes sir!" the young airman responded, with a bit more enthusiasm than was necessary.

I hopped in and he hit the gas.

Named after Ronald Bucca, the New York City Fire Marshal who died in the attacks of September 11, 2001, Camp Bucca was best known to people back home as the infamous destination for prisoners moved from Abu Ghraib after its well-publicized scandal. But most failed to realize that it was an actual military installation boasting an impressive chow hall, a basketball court, a Subway, a barbershop, and a fully functioning gym. Rows and rows of very small, two-person trailers served as lodging for base personnel. Larger single-wide trailers of varying sizes served as offices.

We dipped and dived between rows of trailers and smalls buildings at a steady speed that was almost reckless. The airman transmitted his eagerness through his foot to the gas pedal, as he sped through small openings and thin corridors between structures. I was still unfamiliar with the layout of the base, which probably made the drive feel hastier than it was. I told myself that he knew his way.

I tried to take stock of the landmarks. Through the warm wind and sand pellets whipping across my face, I could make out in front of us a wide stretch of earth with a helicopter pad, sand berms, and concertina wire separating this half of the base from the prison. A now unobstructed view also revealed collapsible wire mesh containers, known as Hesco bastions, lining the roads and the base perimeter. These stood anywhere from ten to fifteen feet high and three to four feet wide, were usually filled with sand or gravel, and were used as blast walls against explosions, rocket attacks, or small-arms fire.

Squinting through the sand and wind, I made mental notes on the locations of port-a-potties. Experience had taught me their value in a deployed environment. The fancy air-conditioned trailers equipped with cramped stalls and tiny urinals, known as Cadillacs, were available only in specific locations. Port-a-potties were as plentiful as they were rotten and sweltering. They symbolized one of our expressions: "Embrace the suck."

I began noticing how we passed Air Force members at almost every turn. I knew there were many stationed at Bucca; after all, we were here to provide them counterintelligence support. But they seemed to be everywhere, and I would later learn that their number nearly exceeded the total Army presence. Between Iraq and Afghanistan, it was obvious American forces were spread dangerously thin. Some soldiers received future deployment orders before even finishing their current deployment. I knew the Army had too many responsibilities and too few people. I hadn't realized it was this bad.

My frustration grew the longer we drove. I had complained more than once in private that Secretary of Defense Donald Rumsfeld seemed unwilling to accept that plans in Iraq were failing. He confused his admiration for the military with its capabilities, and his enthusiasm assumed unrealistic expectations. And now it appeared the largest branch of the U.S. military had too few people to complete the ground mission. The Pentagon had little choice but to fill Army slots with people from other branches. This was sold to Congress as "joint operations." And, as everyone in the military knows, "joint" means working for the Army.

We finally made it to a two-lane gravel road and picked up speed. The skyline opened as we left the congested part of the base. I could now partially make out Camp Bucca's impressive transportation grid. Gravel and dirt roads of all kinds seemed to intersect at multiple points along the path, then trail off and over the horizon. I saw joggers in the distance. "It's great to run the roads," the driver yelled. "You could get lost on this base without them." Indeed, it was flat and directionless. And no one wanted to veer into the wrong part of Bucca.

The prison sat at the end of the central road we now traveled. I could see on our approach that the Theater Internment Facility – the TIF – was a super max prison. Even at a distance it loomed large against the blue, cloudless sky. Cement t-walls scattered outward from the prison. Concertina wire ran across the ground about 30 yards outside the first gate, and more lined the top of the perimeter fence. Elevated guard shacks were situated at several spots along the outer ring. Hundreds of flood lights hung menacingly over the prisoner yards.

The ATV came to a stop, and the airman pointed directly ahead. "You need to go to the security office, right there." Twenty yards ahead, just outside the main entrance to the prison, sat a small plywood guard shack. I nodded and made my way toward it.

The sheltered booth offered immediate relief from the scorching sun. I promptly handed my military I.D. to the young airman manning the guard shack. He looked it over, cross-referenced it with his visitor roster, tersely slid a clipboard with a sign-in sheet across the plywood desk. I signed it and reluctantly surrendered my 9mm Sig Sauer handgun. We carried sidearms everywhere; they were an extension of us. Handing over a piece of me to some grunt manning a shack made me uneasy. But prison procedures trumped my right to carry. My friend, an interrogator at the prison and my escort for the day, ushered me through the fence door.

★

The prison was a city unto itself. I could now see that the once small British POW holding area had grown into a wide expanse of high fences and guard towers overlooking large, crowded compounds of communal tents and cinderblock huts. Activity hummed along like an industrious beehive. The collective noise sounded like a stirring mob. It felt raw. The atmosphere was unencumbered. Energy brimmed on the surface, as if the whole place were one incident away from a riot.

Open to the elements, the entire facility looked like the so-called "yard" of an American prison. Immediately in front of us sat a wide vehicle path that ran between a handful of large, fenced-off internment zones on both sides of the road. It looked severely overcrowded. I could see detainees warehoused in large corrals, sporting bright yellow jumpsuits. Those zones had separate compounds within them that looked to hold hundreds of men. The layers of security put at least two fences and several yards between us and the inmates. Like visiting a zoo, one could simply watch masses of prisoners live out their lives in this fenced-off ecosystem.

I recalled then a childhood neighbor who had owned and rehabilitated Florida panthers. She housed three or four of them in two massive cages in her backyard. Reinforced chain link fences made up the walls. It was a rite of passage for the boys in our neighborhood to sneak onto her property and run between those cages. And those panthers leapt at the cage wall with such ferocity that you would instinctively recoil, only to feel the hot breath of the other panther, its teeth protruding through the gaps in the chain link.

We began our march toward the rear of the prison. Gravel crunched beneath my boots, as my mind absorbed all the sights and sounds. To my left I observed a large group of forty or fifty men huddled near the rows of huts that served as their sleeping quarters. Almost as soon as we passed they came running to the fence, smiling and laughing. We paused momentarily, then soon regretted it. Their collective smell made me shudder.

Between gulps of breath my escort, Joe, explained that this first holding area on the left housed non-violent inmates and petty criminals. That explained their jovial demeanor, he joked. "These knuckleheads get free meals and lodging. They probably steal just so they can get in here," he said in his thick New York accent. He signaled with his thumb to the fenced-off area now behind us, on the right side of the gravel road. "The Shia go over there. We can't house Sunni and Shia together; they'll kill each other."

The gaggle of men at the fence had tripled in size in mere minutes. The stench was overwhelming. With heavy sarcasm, Joe explained: "The military has a brilliant strategy right now. We wrap up any military-age males within so many miles of a blast radius and throw them in here." We all knew that roadside bombs were killing American personnel at alarming rates. And they were nearly impossible to preempt. A terrorist would strap a cell phone to an explosive and rig the combination to detonate when the ringer went off. When a passing convoy neared the device, the bad guys simply made a phone call. Other times, they set the alarm on the phone for a predetermined time, when they knew U.S. forces would be traveling a given stretch of road.

Knowing that terrorists often videotaped attacks or hid close by to ambush surviving soldiers, the military decided to scoop up any twenty-something male who happened to be near the explosion, in hopes of catching the attackers before they could retreat and blend back in with noncombatants. All of those caught were deposited at Camp Bucca until the military could determine culpability.

"They snatch them up and we sort them out," Joe explained.

Some years later, I surprised my wife with a gift by strapping one of her old, unused cell phones to a gift box hidden below our bed. The preset alarm went off as I had planned, leading her to discover the present. When she asked where I had come up with such a creative idea, I told her, "Al Qaeda in Iraq."

Further up the gravel path we passed another yard, where the inmates acted less enthusiastic, even uninterested in our passing. Joe explained how the second section was generally designated for violent criminals, although they were not necessarily part of any terrorist organization. Few came to the fence as we passed.

Before we reached the office building at the end of the road, I looked to my left once more. The last section was much less congested and much calmer than the rest. Several men squatted near the wall of their cinderblock hut. The roof of their hooch provided a short overhang that gave them about three feet of shade. I stopped for a moment and trained my gaze. The glare from the sun made it difficult to see through the darkened area around the hut, so I had to concentrate longer than normal. Two of the men, clearly angered by my staring, immediately stood up and shot back looks of fury. "Oh, tough guys," I chuckled.

Curiously, though, another man maintained eye contact but remained squatting. By now my sight had adjusted to the shade, so

I could see his face clearly. He had a hair-raising look about him, something I felt I recognized instantly, though I had never seen it before, something hauntingly familiar yet utterly new to me. Ego forced me to hold his gaze, and we remained there, eyes locked, for an endless moment. Then one of the detainees who had stood up slowly drew his finger across his throat. I smiled and winked in an ill-advised moment of unnecessary sarcasm. That infuriated him. He began angrily lumbering toward the fence, waving his left hand and yelling something in Arabic. The sun jumped off his bright yellow prison suit as he hastened toward me with a gait suggesting someone who rode horses often. But before he got within arm's reach of the fence, my escort yanked me by the collar and through the open door of the headquarters building.

Suddenly, everything seemed normal again. Computers hummed, coffee brewed, and people chatted over work in a wide, modern space that looked like a massive hotel lobby.

My partner retreated to his desk nearby, with me in tow, and quickly turned his attention to the intelligence reports strewn across his desk. I prodded him gently: "Who was in that last section?" I kept my voice low, not wanting to betray my rookie status to eavesdroppers. I could only hope that my civilian clothes and beard helped obscure my rank and the fact that I was a novice.

"Mostly hardcore Al-Qaeda types." He paused, then finished his thought: "And thanks for stirring them up."

"Oh, that's a friend," I replied as I leaned over to read the papers on his desk. "Inside joke."

"Funny," he said, and handed me a stack of intelligence reports.

CHAPTER 2

A Handful of Hair

I imagine that living in the same house and attending the same church and school since the age of five, along with an international family legacy, stirred in me an urge to explore the unknown.

Growing up, I listened intently to my grandparents' harrowing tales of working as evangelical missionaries in the Dominican Republic. My Cuban grandfather and his American wife from West Virginia proved a dynamic duo in that largely Catholic country. The couple's managing group in the West Indies Mission Board initially contested their marriage as a mixed-race union and a violation of organizational policy. Hilario Diaz, my *abuelo*, told the board without equivocation that God had brought them together and that they would be married. They were, without incident, in 1947.

After the wedding, the two planned to board a ship that would take them to the Dominican Republic. They had arranged to trek overland to their final destination in Azua, a sleepy mountain town about 60 miles west of Santo Domingo. As my mother explained, they arrived at the docks in Cuba to learn that a circus had essentially commandeered the ship. The captain had apparently booted everyone previously listed on the manifest and decided to go with the more lucrative deal offered by circus officials. My *abuelo* vigorously protested the decision. He worked furiously trying to get back on the passenger list, visiting the docks four times over two days, to no avail. Minnie, known for a quiet demeanor that belied an intense wisdom, stepped in and convinced her fiery Cuban husband that perhaps God was trying to tell them something. "You have tried, but the Lord clearly does not want us on that ship," she told him. "Let's just wait and see what He has for us."

The days of haggling led the boat captain and my *abuelo* to a compromise. The skipper agreed to transport their belongings at no cost, but the newlyweds would have to find other means to get themselves to the Dominican Republic. With the agreement in place, the boat disembarked.

The prospects looked bleak. They had little money, and the only belongings in their possession were an accordion, a typewriter, and a small suitcase of clothes. The loss of their seats was devastating financially to the newlyweds, since ships remained the most cost-effective means of travel in those days. But not long after the boat departed, my grandparents received an entirely unexpected financial gift from a church in the United States – just enough to cover the cost of two plane tickets. They immediately secured passage on a local airline.

Days after their arrival in the Dominican Republic my *abuelo* sat reading the newspaper, to learn that a vessel – their vessel, loaded with virtually everything they owned – had sunk somewhere between the two islands. Details were scant, but apparently a storm had caused an open water shipwreck, leaving only a handful of survivors. Circus animals met the fate that might have been my grandparents'.

So, with nothing to receive at port, the couple began their inland trek and arrived in Azua sometime in May 1947, emptyhanded but for the accordion, the suitcase, and the typewriter. I have a picture framed on my wall that captures their arrival. Smiling and wearing their Sunday best, they sit atop two donkeys striding down a dirt path through a shantytown. In their laps lie the accordion and the typewriter.

They would go on to face a hostile Catholic community unwilling to accept the intrusion of evangelicals and, within months of arriving, a campaign of vandalism and harassment led by local bishops. But the threats and violence failed to dissuade Hilario and Minnie.

My grandparents endured and founded several churches in town and in neighboring communities over the next several years. Some of them still stand. They would go on to live through the country's civil war, to share the Gospel with the heinous Dominican dictator Rafael Trujillo, and to leave a legacy of mission work through a street renamed after my *abuelo* in Azua and a body of believers who continue to operate in the country today.

★

My mother recalls moving down from the mountains to the coastal capital of Santo Domingo in 1955, and a slight upgrade from the poverty of small-town life. Any complaints from us children were often met with reminders about her struggles as a missionary child, living with little money, behind mosquito netting, with no air conditioning, and with only leftover encyclopedias for pleasure reading. She would dazzle us with stories of intrigue and memories of war and life under a dictatorship. She often recounted to me the story of the assassination of the infamous Trujillo in 1961 and how Dominicans dared not celebrate, fearing the story was a ploy to lure dissenters into the open. But as confirmation of his death spread through the country, Dominicans took the best revenge they could: They invaded his several upscale homes and stripped them bare. "People were taking whatever they could carry," my mother said.

Several years later the country fell into civil war. The fighting left some streets littered with bodies. My mother grimly recalled the stench that lingered in the air for weeks. During a particularly vicious gun battle near their home, my mother and her younger sister snuck onto the flat rooftop to get a better view of the firefight. Bullets whizzed just above their heads. She would remind us then how my *abuelo* found them and gave them both a reprimand that they would not soon forget. And she laughed at how the small-statured Dominicans were in awe of the large U.S. Marines who had entered the country in 1965 to help restore order. She said the Dominicans couldn't stop talking about those "black and white giants" that marched the streets. In all the turmoil, my *abuelo* told the missionary committee that he and his family were staying put. "We were there to minister to them, and we couldn't run when they needed us most," my mother said with pride.

She would depart her home country in 1968 to attend college in South Carolina. "I cried when I called home for the first several months," she told me, "begging my father to let me study in the DR." Hilario refused. He felt a better education could be had in the United States, and essentially told her to get over it. Homesick, lonely, but with no choice, my mother took on the strict college atmosphere and the new culture with determination and wisdom learned from her mother. Eventually she settled in as a Spanish teacher and counselor at a local Christian school in Tampa, Florida, where she remained for four decades.

My mother found American men uninteresting, lacking the energy and personality of Dominicans. "They dragged their feet when they walked," she said. I laughed every time she said that, but I never dragged my feet in her presence.

My father possessed the spunk she so desperately wanted. A *gringo* with a Latin spirit, she called him. He endured a rough, unloving childhood, followed by equally challenging years as a teenager and young adult. Absent parents and drug abuse pushed my father to the brink. Before disaster struck, he found Jesus Christ and renewed value in life, then became an evangelical minister in Tampa.

He served his time as a hippie but, instead of awaiting a draft notice, he joined up. Asked by the boot camp barber if he wanted to keep his long hair, he quickly agreed – wondering to himself why he could keep his hair and others could not. But before he was able to ask, the barber took a long swipe with the trimmers and handed my father a handful of long red hair.

"Welcome to the Army, Private Grantham."

The Army gave my father the structure and training he needed. He struggled with discipline, earning many hours in cleanup duty. But he went on to serve in the Army's explosive ordnance group, before returning to Tampa to begin training for church ministry. He met my mother, and they were married in 1980.

★

I had the kind of upbringing many boys envied: a father and mother who were present, engaged, and attentive. But it wasn't a childhood without problems. As far back as I can remember, we struggled financially. "If money could talk, it would tell me goodbye," my father would joke. A sister born twelve years after me only added to the challenges facing my parents, who were already raising two rowdy boys. My father sat behind the wheel of a junker for years. No air conditioning, sagging ceiling cover, a radio that at times required we change the station by pinching the antenna. And we rarely ate out. Instead, my father recreated a sandwich shop every Sunday afternoon in our kitchen, designing a massive white fold-out sign on which he scrawled "Subway." My parents provided well with what little they had, and they chose to put what little money was available into things they considered more important, like our Christian education.

Multi-racial and economically diverse, Tampa Baptist Academy was a severely underfunded institution with aging facilities and an infrastructure held together by the fingers of God. Wealthy and poor, black and Asian, brilliant and scholastically challenged, all met under one roof. The cultural training I received just by being there was as beneficial as what I learned in the classroom. And nothing was ever handed to us, except the support of every staff member.

At the close of every school year, when students began celebrating their summer freedom, my parents would begin whispering to each other about the school's fate. My mother fretted about her job. They were always concerned about whether it would exist the following year. I graduated in 2000, and somehow that little institution is still running.

Sports featured prominently at the school, and in my upbringing. My father felt sports would prepare me for life. He took special care to ensure that I saw some larger principle in each situation, rather than mere competition for glory. After a particularly testy basketball game against our hated rival, my father, the assistant coach, grabbed me by the jersey and virtually threw me into the greeting line forming at center court. "Son, you will not be a wimp," he said. "You compose yourself, look them in the eye, and shake their hands. If you can't face adversity here, at a little basketball game, you will lose in life."

Our backyard had two permanent dead spots in the grass, where the pitcher stood and where the catcher squatted. When my 80-year-old great-uncle visited us from Cuba, he saw the yard and knew exactly what it meant. He didn't speak English and didn't need to. He grabbed a glove and planted himself in one of those spots. "*Vamos!*" he yelled – Let's go!

When a hobbled Sid Bream rounded third base on a single by rookie Francisco Cabrera to send the Atlanta Braves back to the World Series in 1992, we were overjoyed. "Braves win, Braves win, Braves win!" still echoes in my mind. That excitement for us was matched only when Tampa Bay finally landed its own major league team in 1998. When Cal Ripken, Jr. broke Lou Gehrig's consecutive games played streak, the announcers remained quiet during the entire celebration. Without looking over, my father said, "The announcers are silent." I hadn't even noticed. Then, without missing a beat, he asked if I listened, or waited to speak.

Sports helped me appreciate giving an effort no matter the outcome, rather than doing nothing out of a fear of failure. I learned quickly that attitude defined my future. Failure is a part of life, and risk is necessary to achieve even the smallest victories. Teamwork, individual commitment, and integrity. Be on time, give your best effort, be teachable, have a passion for something. These things required no skill or talent whatsoever, and often they were the difference between reaching a goal and falling short. And that feeling of reaching a goal proved addictive. I took a competitive attitude to college, deciding to attend the University of South Florida – a public, secular university that would contest, even ridicule, my Biblical beliefs. And I didn't hesitate to choose history as my major.

I chose Medieval History, taught by Dr. Roy Van Neste, for my first history course. I knew nothing of medieval Europe. But moments into the first lecture, I began to question my sanity. Older and small in stature, Roy – as he demanded his students call him – was a quirky free spirit. He laughed often, and he strode around the room as though walking through a meadow on a lazy afternoon. But his supposed carefree personality camouflaged his demanding expectations. He put us at ease within the first few minutes, acting as if he had no intention of making the class difficult. And then, behind his laugh and smile, he read off his professorial manifesto. "Some of you won't pass this class, unfortunately; some of you will fail this course, partly because the education system has let you down, which has left you unprepared for the rigors of college, and partly because this class is incredibly hard, one of the hardest in the department." He gave this little pep talk with a smile, in a jovial manner that made it hard to know whether he was serious.

"So really, it's probably wise for some of you to just drop the course," he went on. "I plan to demand work of you that was demanded of me when I was in high school decades ago." Offering what I would come to know as the chuckle that always preceded bad news, he pulled a stack of papers from his weathered briefcase atop a small metal table and passed the papers around until every student had a copy.

The sheet contained a list of essay questions covering topics like history, society, religion, and economics. "These are questions from my high school exams," he said. He remained quiet for a minute or so, letting it sink in. We read them over in silence. This high school exam was completely over my head. Then Roy broke the silence, launching into a serious lecture about the deterioration of the education system – how the obsession with quotas, sports, and tenure, rather than critical thinking, had caused a decline in what was expected of high school seniors. The whole process had become a sort of government-subsidized racket that was undermining students intellectually.

"Oh, I might add that if I don't offend you this semester, I apologize in advance." He explained how his job was to defy our logic, and our convictions, to ensure that we thought critically about everything. "Whatever you believe, I hope I upset you at some point," he said. "If not, I have failed at my job." This was exactly what I had signed up for. And, ironically enough, I had not expected such forthrightness. Roy provided much-needed encouragement the second day of class – a class that was about half the size of the first day. "I reviewed the roster and saw several freshmen and sophomores in this class," he said to start. I braced myself for another warning about the rigors that accompanied his assignments. Instead, he offered encouraging insight. "I know you are smart," he said. "Why? Because the bright ones always choose a class that is completely unfamiliar."

Interesting observation, I thought. Never had I considered myself a bright student, just a hard worker. But Roy's comment gave me pause. He continued: "The average students generally gravitate toward topics that are familiar, topics in which they already have a working knowledge." That truth registered with me. It was for that very reason that I had avoided American history, after years of it in high school.

"Not to discount the intellect of those not in here," said Roy, "but I know for certain that those of you still in this class are some of the intelligent ones."

That was all I needed to hear. This class would be my crowning achievement. I would conquer this course with a grade that would surprise even Roy. But the first paper I submitted was returned with no grade, only the comment: "Come see me."

In his cramped office in the windowless basement of the humanities building, Roy wasted no time. "I know you're a good student," he said. "Private school students usually are. The smaller classes allow for more one-on-one attention." I waited for some slight about the school's Christian mission. But nothing like that was forthcoming.

Instead he said, "And I know you are a good student because you sit in the front row, you're attentive, and you're on time to class." It surprised me that he would remember those details. "You always take a seat right in front of me. You take notes, you ask questions."

His observations caught me off guard. But the compliments were short-lived. "You know how to write, but your organization is terrible, really incoherent. And you clearly have little understanding of how to discuss history." Stung both by his candor and by what I knew was a correct assessment, I sat in silence.

He then offered a strategy for writing well, breaking down a method for constructing prose and organizing my research for the collegiate level, and explaining how to develop a robust thesis. In about twenty minutes, he revolutionized how I studied and how I approached the craft of writing, and transformed how I understood education.

By semester's end I had achieved a C, by far the lowest letter grade of my entire collegiate career. Roy's class proved to be the anchor that dragged down my grade point average each and every year. Fractions of a point separated me from superior academic levels and the top echelon of college awards. But the grade didn't reflect how much I learned.

Throughout the semester, Roy offered historical training, academic insight, and personal advice that I carry with me to this day. Yes, he said things I did not agree with – and yes, he angered me, as he had promised he would. But his approach forced me to examine my beliefs, strengthen my position. Entering active duty after college I felt ready to tackle the next challenge. Quickly I learned that the principles and wisdom Roy instilled in me would prove the difference between life and death – both stateside and in a deployed environment.

CHAPTER 3

AMERICA THE BEAUTIFUL

It was nearly summer, and the musty, humid odor of the windowless room in the old buildings at the university proved exceptionally strong. Despite its unwelcoming appearance, I had felt secure on the bottom floor of those bunkers during the attacks of September 11, 2001. We had arrived at class about an hour after the towers fell. The atmosphere was quiet, tense. Before dismissing the class, the professor posed a question: "Who do you think did this?" Having read much on war and intelligence, even at that age, it seemed clear to me this was probably Middle Eastern terrorists, not domestic terrorists. These guys died for their cause. Timothy McVeigh and others like him rarely did. I blurted out, "I think it's radical Muslims from overseas." Two fellow students reflexively chastised me for making it a "Muslim thing" and ignoring domestic terrorists. I snapped back that it was evidence, not prejudice. The professor quickly dismissed the class.

I now sat in a similar classroom two years later, awaiting my first assignment as an Air Force officer. Graduation was scheduled for the following week, and today we received our duty station location, having filled out the dreaded "base preference" sheet months ago. My top three choices were international locations. The remaining two were bases in the Northeast. I surmised that, if the Air Force refused my requests for Europe or Asia, I could still enjoy baseball at the most iconic stadiums in the game.

Some cadets had inklings of their destinations. One friend, an avid hunter, wanted open spaces and plenty of game to track. Naturally, his first choice was Minot Air Force Base in North Dakota. The frozen tundra was judged the least desirable across the Air Force, so he knew his request would not be denied.

Some cadets were chatting with friends, others fidgeted. I was anxious that the Pentagon might assign me to MacDill Air Force Base in Tampa – too close to home. My ROTC commander, Colonel Johnson, had succumbed to a fit of laughter when I expressed my worry to him. "Cadet Grantham, if you didn't request it, you won't get it," he assured me. "Everyone wants to come to Florida." I was relieved and a little embarrassed. Yet now I sat quietly, still nervous that my quest to see the world might begin a mere twenty minutes from home.

As it turned out, the Pentagon had assigned me to Ellsworth Air Force Base in South Dakota. My stomach sank. I was once told the base preference sheet was a joke the Air Force played on unsuspecting cadets. "You'll get sent where they want you. Your opinion doesn't matter," one Air Force veteran told me. I now believed it.

I came home dejected. My parents found it funny, until they realized how far away I would be. None of us knew how to plan for the big sky country. I had seen snow only once in my life. It would be like traveling to a foreign country. We shared a common language with the locals, but that was it.

One of my parents remembered then that my uncle, who had served in the Air Force for twenty years, had been stationed somewhere in the snow-blanketed American West. I phoned Uncle Ralph at his home in West Virginia and, to my delight, those infamous stories of ice-covered runways and massive snow drifts came from his time at Ellsworth Air Force Base. We spoke for hours as he enlightened me on the rigors of life in the Dakotas. He told me where to live and offered tips on which vehicle to purchase – a process that forced me to shop out of state, since the inventory of four-wheel drive trucks in Florida was quite thin. A month later I found myself loading belongings into a new four-wheel drive truck and hugging my proud but grieving parents. My sister waited nearby for me to drive her to summer camp. Our twelve-year age difference made a special bond between us. I had proudly helped watch over her from birth, and now I was devastated to think that she would enter middle school and high school without me there. When she began to cry at the dropoff, so did I.

The government gave me six days to travel from Tampa to Rapid City, and I took full advantage. My slow trot across country proved an eye-opening experience full of unfamiliar and fascinating scenery and culture, from Atlanta to Nashville, St. Louis, and Sioux City. The legend and history of the Grand Ole Opry; the blue-collar spirit of a small industrial town in Iowa; immense acreages of massive corn stalks; the size and strength of the Mississippi River. And a visit to the old Busch Stadium in St. Louis. The nation I had sworn to protect held vast beauties and a variety of culture that I had never anticipated.

It was the four-hour drive from one end of South Dakota to the other that truly caught my imagination. In my mind's eye, Sioux Indians on horseback pursued herds of buffalo across the state's rolling plains. The tall grass bent with the raging wind. The sky seemed to go on forever. I was reminded how far I was from Florida when a Native American convenience store owner asked skeptically if Floridians could pull oranges from the tree and eat them. Having grown up doing just that, I was a bit perplexed. "You sure can," I told him. He shook his head in amazement.

My daydreams faded as urban sprawl slowly appeared over the horizon. Within seconds, I could make out the air base from among the neighborhoods scattered around it. Distance plays tricks on depth perception in big sky country. Ellsworth seemed to be no more than a twenty-minute drive from where I sat, but another forty-five minutes of driving and I still hadn't arrived. My uncle's words rang in my ears: "You can see things coming for hours." He had also told me there was a pretty girl behind every tree in South Dakota. I now knew that to be a joke, as I surveyed the grassy countryside.

★

My new supervisor met me at the gate. An exceedingly kind and energetic man, Terry Krebs spent the next several days walking me through the in-processing procedure – something that required stamina and patience. New housing. Financial arrangements. Weapons storage. Medical evaluations. Office assignments. Security badges. Thankfully, Krebs had over two decades in the military and knew precisely how to handle it. But when he learned that I owned only a windbreaker, he whisked me away to a local sporting goods store, where he provided an on-the-spot tutorial in all kinds of cold weather clothing. I was amazed at the effort it took to clothe oneself in winter.

On our return to base, Krebs said that he had personally requested me. I wasn't sure why he would have. I had no history with him or with Air Force special investigations, nor did my skill set read as any more attractive than any of the other college graduates who entered the command at the same time. He continued without much of a pause, saying that another agent with the exact same name had recently retired from the command. This other David Grantham had been a popular, well-respected figure. "I wanted the new guy with the same name," he said. "Oh, and the only reason you got in was because the selection board thought you were Grantham's son." He paused as if serious, then jabbed an elbow into my ribs. But I hadn't considered that my selection for the most competitive career field in the Air Force might have been dumb luck. The thought gnawed at me. In the meantime, I received phone calls weekly, from rather important people across the command, who thought I was the original David Grantham returned as a civilian. I made some important friends early on.

★

Just two months in South Dakota, and I was on the road again. My orders now had me returning to the southern states for nine months of training split between Air Force officer instruction in Alabama and much longer, far more grueling investigator training at the Federal Law Enforcement Training Center in Glynco, Georgia, a small coastal town just north of the Florida line. My schedule there involved both classroom lessons and field application. Some days were devoted to dissecting legal doctrine and drafting search warrants, others to high-speed driving and firearms instruction. Recent graduates bemoaned the seemingly infinite firearms exercises and told me I would tire of the shooting. What I initially found hard to believe proved true. Hours upon hours of training made the gun feel like an extension of you. Every graduate understood and deeply appreciated the enormous responsibility that came with this newfound authority. And although I earned a top marksmen's award, after graduating the academy I would have been fully content never to fire a weapon again.

We spent an equal amount of time learning the art of source handling: cultivating relationships with unsavory people without compromising our own principles, working a well-meaning source without growing too fond of them, building trust with a potential asset, eliciting information and making "the pitch," the official request of someone to be an informant. Quickly I learned how to motivate informants, and how to elicit information through conversation. "It's conversation with a purpose," our instructor said during an intense seminar on intelligence collection. I learned quickly that, done well, elicitation required an appreciation for the workings of the human psyche. It also demanded integrity, selflessness, and teamwork, because without those elements such deceptive behavior can become part of one's identity.

★

In March 2005 I returned to Ellsworth. I was enjoying the criminal investigations, but the harrowing stories coming out of Iraq and Afghanistan left me wanting more. I angled for an opportunity to join the fight, knowing full well that deployment would be unlikely for a junior officer like myself. So, I plotted a course to overcome it.

The mission in war for OSI is counterintelligence, and it required that agents perfect their ability to handle sources. In hopes that successful work would earn me a ticket to the Middle East, I made cultivating informants a top priority. I took great care to assess my performance, I sought out criticism, I begged for specialized training. My tenacity began to pay off, as my informants helped uncover a major drug ring on base, at least by Rapid City standards, resulting in several apprehensions. Those efforts so early in my career, along with two other important cases of child abuse and fraud, earned me the Rookie of the Quarter award. Krebs soon after entrusted me with the office of counterintelligence, a major goal of mine, partly because that experience would qualify me for the warzone. It was only a one-man shop, though; South Dakota was not a beacon for foreign intelligence.

My attention was naturally drawn to the B-1 bombers stationed at Ellsworth. And some further probing taught me that these pilots flew all over the world, often to hotspots of foreign spies. Their travels and the follow-on debriefings offered me the opportunity to hone my skills at intelligence collection and analysis. I was able to jump headfirst into the intense process and, to my amazement, produce some worthwhile intelligence reports on spy tactics used to collect information on America's most prized bomber. The work not only improved my confidence; it also gave me a taste for the intrigue that came with the addictive secret world of spy craft.

Soon after I was given the reins to the CI shop, a colleague returned from Iraq with stories of intrigue. His work with informants had uncovered a major weapons stash. Some Al Qaeda members were found and arrested. Threats to the base stopped cold. He went on and on. As luck would have it, he moved on to his new duty station at OSI command headquarters near Washington, D.C., with full knowledge that I wanted to be on the next plane out.

"They are hesitant because of your rank," he told me over the phone a few months later. I was thrilled to hear that he had already begun to float my name. "Let me continue to ask around and see what we can do." Since we deployed one at a time, not in units, I had to wait for a spot to open. The process was competitive, and sometimes it took a measure of luck to find yourself with orders to go.

As fate would have it, only a month later we all were made aware that a fellow agent in our office had been disqualified from a deployment due to injury, and OSI needed someone from our office quickly to fill the spot. My friend at headquarters once again tossed my name out there as a replacement. A debate ensued among the higher-ups, some of whom were unconvinced of my abilities and wondered aloud about my maturity. Luckily, I had a resource in the building, an advocate who knew me very well.

★

"We got the green light," he said bluntly. "Just keep your mouth shut until you hear officially." My Beltway buddy gave me the good news on a Friday afternoon, just before a long holiday weekend. I was ecstatic. Keeping the information under wraps throughout the three-day break would be hard. But before I could ponder that possibility, Krebs was in my office. He grabbed one of the two cushioned chairs along the wall and pulled it directly in front of the desk. Without a word he sat down. He began to fidget with the pen on the edge of my desk. A few seconds later, he returned the pen to its resting position and slowly began to smooth out his tie. The long pause felt like an eternity.

Just as the silence started feeling awkward, Krebs looked up at me with a big grin. "You're going to war, Cupcake," using the stupid nickname I had fallen backwards into at Ellsworth. A smile escaped from my otherwise stoic face. "You've earned this," he said. "I want to keep you here, but I know this is important for your career." He looked down at the desk and, without lifting his head, said with an unusual level of genuineness, "You are like a little brother to me, so don't get yourself killed." No sarcasm. No mockery. No chitchat. Finally, our eyes met again. "And if I get one phone call that starts with "Frickin' Grantham …" I smiled and nodded. I wanted to say thank you, but that didn't seem like the appropriate response. And I didn't have it in me to be sentimental. So, I just muttered, "Yes, sir." With that, he stood up and walked out.

CHAPTER 4

The Crown

I deplaned onto the tarmac at Kuwait International Airport full of nervous energy. Beyond the bright lights that bathed the runway directly around us, it was pitch black. Nothing moved in the quiet terminal at the end of the strip. Our midnight arrival came without fanfare. I scanned the small crowd of onlookers gathered to escort new arrivals, and I found myself inspecting the darkened horizon for threats. I tried to mimic the calm of those I saw waiting for me, but still I looked around uneasily, as if we had landed under a sniper warning. "Welcome, Grantham," my new boss said as I descended the stairs that extended from the plane. "Relax," he said gently, as if reading my body language. "Grab your bags and let's get to the hotel."

Those at OSI headquarters felt Kuwait would be a better location to prepare a young officer for the rigors of counterintelligence in a foreign environment. I had already come off the plane ready to engage the enemy, so perhaps they were right.

"How was the flight?" the boss asked, turning from the front seat to face me. The SUV we were in flew down a deserted highway leading away from the airport. "Good," I said distractedly, as I tried to make out the surroundings in the darkness. Light posts every 50 yards provided brief glimpses into the night. "We are headed to the hotel," he said, turning back around to face the front. "I think you will find your lodging worthwhile." He laughed. I nodded, not wanting to admit that I had no idea what he meant.

Until that moment, the upscale aspect of my deployment had been unknown to me. My deployment began like few others: living in the Crown Plaza, a five-star hotel in downtown Kuwait City that boasted a state-of-the-art hookah lounge, a modern gymnasium, and six restaurants, one of them a premier steak place with incredibly attractive Filipina waitresses. We passed through security, and the boss directed me to the front desk. The lights in the massive lobby bounced blindingly off the ornate decorations. The opulence of the place was like nothing I had ever experienced. Perhaps it was the jet lag too. My sensory system was severely over-stimulated, and the need for rest suddenly hit me.

★

"Here's the deal," a veteran OSI officer explained about a week later. "The first agents that came to Kuwait got to stay in hotels because the base was too far away and source meetings lasted until three or four in the morning." I sipped my soft drink and nodded. Having already experienced some of those late-night meetings with government informants, I could appreciate the dilemma. More than once, I had stumbled into my room in the early morning, only to peel myself out of bed a few hours later to report to work at Ali Al Salem Air Base, where our office was, an hour's drive outside the city.

"Still, seems a bit lavish," I said, more reluctant than argumentative.

"Most of our work is at night in the city anyway," the OSI veteran responded. "Sources pray five times a day and have their afternoon siestas, so evening is when we get things done. Lucky for us, hotels in Kuwait are either security risks or five-star locations, and the government won't approve roach motels." It all really became permanent, he explained, when the Air Force decided that airmen stationed at Kuwait International Airport needed to be lodged in downtown hotels. The Pentagon had to accommodate Air Force personnel responsible for the arrival and departure of American military aircraft, and there was no base nearby to house them. On top of that, the Kuwaiti government was dead set against building lodging that "looked permanent," refusing to build barracks at the airport. Those Air Force personnel stayed in hotels of such luxury because there were no other options. "So us counterintelligence folks naturally followed."

He didn't mention in that conversation, but I learned through experience, how operational security issues also factored into our hotel arrangements. The distance and open roads between the hotel and the base gave anyone watching us plenty of time to react and establish surveillance. If we lived on base, following us would lead anyone right to our source. The hotel gave us more flexibility and more cover. We could leave on foot. In a vehicle. In a taxi.

Lodging downtown had other, less obvious perks, which I hadn't considered. We began regularly venturing into the city for dinner at my request. I wanted to experience Kuwait. So my linguist, a 6'6" Sudanese man with a big smile and hearty laugh named Ramel, often escorted me out to local dives in the evening. I would creep slowly down the teeming streets of Kuwait City at night, hands gripping the wheel of our SUV, feverishly trying to avoid pedestrians and vehicles that always seemed mere inches from us. The jammed roadways left room for a prayer and little else. No one showed the angst I felt, though. Locals routinely pushed carts between moving vehicles without care. Motorbikes wormed their way through crowded streets with ease. Pedestrians wandered in groups across busy intersections. But the biggest hassle was parking. It wasn't simply finding a parking spot. The window washers were always out in force, swarming our vehicle before we came to a complete stop. The first time I heard Ramel rebuke these young entrepreneurs, I thought it harsh. But I quickly learned that, without a stern refusal, these ambitious youths would wash the windows and the car without permission, then demand payment.

One evening, we stumbled into a small restaurant that boasted an unforgettable fish and rice meal. I had almost perfected the local technique of eating rice by hand, mushing clumps together with one hand and using my thumb to flick the ball of rice into my mouth. I had become so good at it that the cook stepped out from the kitchen to see for himself this skinny, young white American and his very tall Sudanese sidekick eating like locals.

★

My first major security briefing came about three weeks after settling in. We traveled to a local Army base where all the major American intelligence representatives in country had congregated to hear the latest intelligence and to deconflict sources. A local Army colonel began with an overview, reminding newcomers like myself that as a small, cosmopolitan country which shared its northern border with Iraq, Kuwait acted as the staging ground for many forces headed north. "Logistical lines live and die in Kuwait," he said. Indeed, some of the most critical convoys that sustained the war effort originated in the filthy rich oil kingdom.

I learned through the briefing that this modern city, with all the amenities of the Western world – malls, fast cars, fine dining – also played host to a celebrity list of espionage professionals from countries including Russia and Iran, there to monitor and collect information on U.S. troop levels. Adding to the interests of those adversarial governments obsessed with U.S. activities were allied forces on our bases, like the Japanese, Australians, and British. And Kuwait was also crowded with U.S. intelligence agencies. "Look around. We got Army, Navy, Air Force, and other secret groups here." I was only the latest arrival in a long-running and crowded game.

Before the colonel yielded the floor to a Navy intelligence analyst, he reminded everyone to deconflict informants. "Make sure you aren't running the same sources." The September 11 attacks had made it painfully obvious that it was crucial for intelligence to be collected and processed in a coordinated fashion. But I found myself still learning how to operate foreign sources. Foreign assets had to be understood – their personalities, motivations, and access – all within the context of their culture. And that was something that came only with experience.

The Navy analyst transitioned to discussing threats to the local seaports and shipping lanes. I listened intently and reflected on my genuine interest in foreign cultures as a point of vulnerability. I needed to be wary of growing too fond of sources in my quest to enjoy and understand the culture. "Misplaced trust can blind a collector. Emotional attachment can be one's undoing," one instructor had said. Relationships develop naturally when working in high-stress situations, he would say. It breeds camaraderie. "But remember that everyone is a potential source." Anyone who handles sources comes to this cold realization at some point, whether through deadly consequences or otherwise, and it's important to maintain a proper emotional detachment.

★

"Dinner tonight with the Japanese," the commander yelled as he darted down the hallway of our small office. I had returned minutes earlier from my briefing and realized the day would continue. A few heads poked out from office doorways, hoping to get more information from the boss before he turned the corner and disappeared through the front door.

A little before 6:00 p.m., the team gathered in the common area of the office as we readied ourselves to walk over to the Japanese chow hall. "Do we know anything about these guys?" someone asked. The team clearly had concerns about what we were about to do. Counterintelligence agents see spies everywhere, and we had little time to research the Japanese troops we would encounter. I pushed my concerns out of my mind as the boss walked in. I rationalized that if he set it up, he knew what was going on. I thought myself too new to question it.

We arrived minutes later to find half a dozen Japanese soldiers sporting wide smiles. They all wore desert camouflage uniforms not unlike those of U.S. troops, and they escorted us into the building and seated us at a long table already set with plates, bowls, and drinks. They bowed repeatedly. Their hospitality was so deferential that it made me uncomfortable. But I went with it.

I had Japanese troops to my left and right, neither of whom spoke English. We communicated through hand signals and through the broken English of a Japanese officer at the end of the table. This cycle of awkward communication continued for several minutes. The chef, a diminutive older man, took everyone's attention as he entered with a massive tray of food. "I love to try new food," I said to my partner, who sat directly across the table. No sooner had those words left my mouth than the chef plopped down on my plate a long slab of eel. I froze, hoping no one saw the look of horror on my face.

The eel looked as if it had only moments earlier been pulled from an aquarium, skinned, and plated. No seasoning. No oven or stovetop markings. Just raw seafood. I stiffened and tried to conceal my unease. Seafood was not my favorite dish anyway, raw fish even less so. I was torn between guilt and nausea. But, I pointed out to myself, I had no choice but to eat this.

I had not forgotten my childhood techniques for finishing unpleasant meals. I gathered the rice by the bowlful and began soaking the eel in soy sauce. The Japanese soldier to my right shook his head as he watched me apply liberal portions of rice and sauce to my plate. "Sauce too much salt. Not good for you," he said disapprovingly. He met my shrug with a smile. I pretended not to notice how, moments earlier, he had feigned complete ignorance of English. This one is probably a spy, I thought.

I went to work on the eel, cringing with every bite. The soy sauce and rice did little to mask the intense fish taste. Ten minutes later I was finished and sat amazed at what I had accomplished. My chest puffed out a little as I watched other Americans struggle to finish and basked in the approving glances of the Japanese.

Then the chef noticed my empty plate. I returned his look with a proud smile. His eyes widened. A large grin formed. Then he retreated to the kitchen without a word.

Moments later he returned with a second piece of eel, larger than the first. I squirmed in my seat, as the eel slid from his spatula onto my plate with a slimly thud.

I looked up to see an amused officemate wink. It dawned on me that some knew to leave a little on their plate to avoid second helpings. I had cleaned my plate. Just then, a belch brought with it a waft of eel and nearly triggered my gag reflex.

All I could do was wave them off. "I can't eat another bite," I said, as apologetically as I could manage. I offered seated bows, in a desperate attempt to rid myself of the food. The chef protested a few more times, before agreeing to take back the eel. It seemed he took no offense. He motioned congratulations to me for finishing the first eel.

Relieved, I stood to join others who had begun congregating along a far wall by the entrance. "Picture time," said the boss. We gathered against the tan backdrop with some of the Japanese troops. About five cameras were set out in an orderly fashion on a small table. One soldier insisted on taking all the pictures and hurriedly pushed others into the frame. He clearly did not want to be photographed. I figured he was an intelligence collector. As he grabbed my camera, I asked the chef, who was wiping down the tables, to take the photo. I now had a small level of rapport with the chef, so I thought he might oblige me. "You," I said, smiling and pointing. "Come on over here." I waved for the soldier to join the picture. He refused multiple times, before finally relenting. He was not pleased.

The photo captured an annoyed Japanese soldier who only barely made the frame.

Immediately after our return, the boss was on the phone with headquarters. We all loitered just inside his office door while he chatted away. The office air was noticeably cooler at night. It was around 10:00 p.m., and we still had a long drive ahead of us. I fidgeted with my shirt button, hoping the conversation would end quickly so we could go.

"Yes sir, with the Japanese." He paused. "Just now. We left about five minutes ago." A concerned look crossed his face. "It seemed to go well, yes." A long pause. A few more words of agreement that sounded almost apologetic, then he hung up the phone.

"Headquarters was not too pleased about us meeting with the Japanese," he said, clearly disappointed in himself. He hung his head and put his hands on his hips. "We should have been more prepared."

Several of us nodded in agreement. I kicked myself for not expressing my concerns earlier. Intelligence can be an exhausting, never-ending game of brinkmanship. But you can never let your guard down.

Finishing his thought, he said, "I was told they are some of the most aggressive collectors here." That was a subtle rebuke – HQ's way of saying there was now probably a folder in Tokyo identifying each of us as a U.S. intelligence collector. I took comfort in knowing that mine would be partially wrong – it would say that I was fond of eel.

★

In bed that night in my hotel room, I clicked on the television and turned to ESPN, the only station in English worth watching. The connection wasn't great, but I wasn't about to complain about having a TV. Or a private room. I was spoiled to the point of embarrassment. I fell into a sound sleep for the first time in days, only to be startled awake by a thunderous commotion below my 6th floor window. I peered past the curtain expecting to see an attack in progress, perhaps AK-wielding terrorists. Instead I saw hundreds of men running through the streets, jubilant. I called down to my contact in the security room. "Egypt just won the Africa Cup," he said. I learned that night that Kuwait had a large international labor force and, apparently, Egyptian workers had congregated near the hotel to watch the final match. My frayed nerves calmed, I dozed off again. The 4:00 a.m. call to prayer at the mosque across the street would sound soon.

CHAPTER 5

Death's Opening

My computer monitors flickered to life in the early morning hours of February 22, 2006. I was now settling into my role as an intelligence officer in Kuwait. For a deployed environment, the office space was incredibly clean and uncomfortably quiet. It felt like an office stateside, save for my desert boots and war weapons. The world outside was unforgiving, with all the elements of a warzone: cement barriers, loud, filthy transport trucks, weapons, tents. Inside were sterile white walls, nice office furniture, carpeted floors, modern electronics beeping and moaning at regular intervals. The only hint we were not stateside was the fine sand that snuck through cracks and gaps in the wall and left a light film over everything. So far, I was being thwarted in my hope of ditching the creature comforts of home for a rough and rugged lifestyle in a dangerous foreign land.

Still half asleep and not a coffee drinker, I halfheartedly scanned the cache of intelligence reports from overnight. I analyzed, and over-analyzed, all possible angles, the source's intentions, the information. I considered possible locations for the next meeting, based partly on the surveillance opportunities of the adversary. I mulled over the conversations about Iranian military units in Iraq and their connection to the brewing civil war between Sunnis and Shias. Iran was Shia after all, and quite interested in covertly disrupting American efforts. Al Qaeda was Sunni, with its own agenda of bloodletting. The last issue was of particular concern, because it threatened to undermine U.S. credibility among a frightened population and any hope for stability in Iraq. Few fully grasped the dynamics at work within the domestic strife and even fewer knew how to address it.

I had yet to truly focus on the stories in front of me when intelligence reports began to trickle, then flood in, followed by foreign news coverage of a massive explosion that had ripped through the venerated al-Askari mosque in Samarra, an ancient city on the eastern bank of the Tigris River about 75 miles north of Baghdad. The blast had shaken buildings for miles, a reporter said, sending all manner of debris spewing skyward like ash from a volcano. An unrelenting fire prevented responders from advancing onto the site for hours. Newscasts recounted how residents had been forced to watch in horror as the signature golden dome of their revered ancient mosque, one of the holiest shrines in Shia Islam, sustained irreparable damage. Looking on in horror as video clips replayed the devastation, I knew the unprecedented bombing of one of Iraq's oldest and holiest cities signaled a frightening escalation in the burgeoning civil war. The lack of injuries did little to stem the tide of anger among Shia everywhere. The strike represented an unmistakable shift in insurgent tactics and sent a charge through all levels of Iraqi society.

News stations in the United States picked up the story soon after. It was an unusual sensation to see news of an incident only miles down the road captured and rebroadcast thousands of miles away in no time at all. Protesting mobs swarmed the streets, calling for revenge and setting fire to Sunni mosques. The violence quickly engulfed major cities including Basra, just one hour north of Camp Bucca. The president of Iraq called for three days of mourning. President Bush publicly offered his condolences. The Iranian supreme leader, Ayatollah Ali Khamenei, capitalized on the tragedy by accusing American intelligence services and Israel of perpetrating it. Iranian sympathizer and famous Shia cleric Ayatollah Ali al-Sistani followed suit: "If the government's security forces cannot provide the necessary protection, the believers will do it."

The blast site in southern Iraq fell into our jurisdiction, so I readied some initial analysis based on intelligence reporting. The evidence pointed to a Sunni insurgent operation, most likely carried out by Al-Qaeda in Iraq. President Bush would soon admit as much in a public statement. We knew AQI's leader, Abu Musab Al-Zarqawi, wanted to spur the Shias into open conflict with Sunnis. Meanwhile, the ayatollahs hoped this would push local Shias into the waiting arms of Iran. Despite some disagreements within the immediate analysis, everyone agreed that the audacious move against a Shia landmark had few analogues in recent history. The attack had crossed the Rubicon of sectarian violence. All knew the significance, but no one yet knew what the backlash would be.

★

Jordanian-born Zarqawi had joined the Mujahedeen's war against the Soviets in Afghanistan in the late 1980s, but he missed the opportunity to fight. There he met Osama bin Laden, then returned to Jordan hardened, with plans to create his own terrorist organization. His arrest soon after for weapons possession landed him in prison, where he earned a reputation as a violent enforcer and cemented his militant convictions under the influence of popular jihadist imams with whom he was incarcerated.

After his release he fled to Afghanistan, where he first received funding from bin Laden to resurrect another branch of Al Qaeda. After the U.S. invasion of Afghanistan, he fought and then left to take up residence in Iraq. Colin Powell famously used his presence in the country as evidence of Saddam's connection to Al Qaeda. The scant linkage left something to be desired, but Powell's reference to Zarqawi propelled him to stardom, and he used his newly formed wing of Al Qaeda in Iraq to prepare for the coming U.S. invasion, which took place in March 2003.

★

Over the next week I monitored after-action reports. Retaliation strikes continued, sporadically for now, across the country. Revenge was coming to mean the massacre of fellow believers. Zarqawi, a

Sunni, maintained his focus on Iraq's Shia majority and followed the mosque bombing with a campaign of bloodletting: a harbinger of the time to come, when the organization would rebrand itself as ISIS.

I soaked up as much information as possible. But it became clear that policymakers feared this attack and the subsequent violence would reinforce the image of a failed American effort. Zarqawi was quickly and successfully pitting Sunni against Shia, and both against the U.S. military. We watched helplessly as Baathists and former Saddam loyalists swelled the ranks of Al-Qaeda in Iraq, giving it experience and a new resolve to defeat the occupying American forces. Things appeared to be coming undone. We couldn't act fast enough.

We responded to requests for information by working our sources overtime. Our Kuwaiti contacts knew we needed more information on Al Qaeda: numbers of fighters, funding possibilities in Kuwait, strategic insight into Zarqawi's intentions. We met regularly to gain a better understanding of what was likely to transpire. We decided to go to the deserts of northeastern Kuwait, near the Iraqi border and a stone's throw from Iran. The police at the far-flung outpost routinely had intriguing information, if you could get there.

The doublewide trailer served as the police station in this forgotten corner of the Kuwaiti desert. "How can you even tell where we're at?" I asked our linguist as we pulled up. He let a toothy grin creep out. "My man, you have to do it a few times before you know the sands." The shapes of the land and paths in the sands revealed themselves only to the experienced. "That is Iran," he said jabbing the air. I peered through our dusty windshield at a large expanse that looked to be only ten miles or so away. But the desert, like the South Dakota landscape, plays tricks on one's sense of distance.

We entered unannounced, and traditional Arab hospitality took over. The four officers jumped to their feet and promptly told us to take seats on the floor around a massive bowl filled with steaming rice and chicken. This was typical for communal meals. Several chickens are placed in the middle of a five-foot-wide metal bowl, and loads of rice are spread around the outside. Each person commandeers a spot along the rim.

"Thank you," I said reluctantly. I was hungry. My female partners declined.

"What have you heard about Iran?" I asked the officers as I cleared away half-eaten chicken from my new spot at the bowl. "They move back and forth over the border, or that's what we hear." I shoveled a handful of food into my mouth. My linguist ad libbed, "You haven't seen it?" They shook their heads in unison.

We asked about aircraft, weapons, and people. They didn't have anything specific, or at least nothing worth reporting. A couple of them looked like they were holding back. My linguist later told me he got the same impression. But they weren't going to speak up in a large group, and we couldn't get them alone. We filed that away for future meetings.

"What about the mosque? We hear it was probably Zarqawi." Several of them nodded. "Be careful with the Shia," one officer said. "They may seem to be not as violent as Sunnis, but they can be." That thought stuck with me. So much attention was given to the Sunni outfits, like Al Qaeda, that the Shia were often represented only as victims.

"What will happen next?" I asked. But they said what we already knew: that bombing one of the holiest sites in Shia Islam was a point of no return. The powder keg was primed and ready to blow.

We loaded back up. I was content with my full stomach. Ironically, the agents who had not eaten would feel unwell in the coming days. I looked forward to my next meal.

On our way out of the desert, we stopped by a Bedouin tent. Bedouins have been wandering nomads since the days of the Bible. They showered us with dates and eggplant.

We were mindful to remove our shoes while in their sizable tent. "Relax, don't worry," one of them said, after I apologized for showing the soles of my feet. I had instinctively propped them up without thinking as we lounged on their thick, ornate rugs. "You are not Arab, so we cannot expect you to do what we do." The interaction was enjoyable, but it yielded nothing in the way of intelligence.

We returned to the office after a lengthy drive back. And before I could get my head around the issue in front of me, another assignment came down: the one I had hoped for, and one I wished had not become available.

★

"Grantham! Get in here." The command was punctuated by profanity, which made me move a little faster than usual. I pushed away from my desk as if in a fire drill, sprang from my seat, hurriedly made my way down the short, narrow hallway to the corner office. Leaning my head just past the threshold to the commander's office, I quipped, "It was Erin, boss." Erin didn't even look away from her computer screen. A broad smile followed by a light chuckle was all she offered.

Erin was two years my senior and already had some experience as an agent. She was a short, attractive officer with a petite frame and winsome voice, whose dark hair was often pulled back into a small, tight bun. A Maryland native of mixed Asian and Italian ancestry, she had a calm demeanor accompanied by a bright smile that seemed to reach her eyes, all of which belied her incredible toughness.

Her mother had left her and her younger brother at an early age, leaving her dad to raise the two children alone. Erin had suffered a near fatal vehicle accident during a summer break in college. She told me that what had scared her was not the grueling rehab, but the possibility of losing her highly competitive scholarship to the Air Force Academy. So, she had rehabbed all summer and hoped the school would not dismiss her when classes resumed in the fall. But the Pentagon's convening board determined that her injuries were too severe for continued enrollment at the Academy, and the government had rescinded her scholarship.

Interestingly, Erin had returned to college in the fall unaware that a decision had been reached, and the Academy had no idea an accident had even occurred. She asked no questions and nursed her wounds in private. The heavy scarring traveled the length of her leg and up to her hips and abdomen; her clothing thankfully hid the worst of it. Then, one day, she found herself summoned to the admissions officer.

"Have a seat, Berlow," he commanded. "You have something you want to share with me?" Feeling there was no point in trying to bluff, she went into painstaking detail about her accident, her recovery, and her return to school. She explained how she had performed painful rehab and wound care in private for months. She fully expected to be booted from the Academy.

The commander listened intently. Then he said: "Well, you almost died, made it back to school, and carried out all the routines as if you were a hundred percent. No one has even noticed. So why would we kick you out now?"

Erin graduated with her class, not only beating the rigors of a military academy with life-altering injuries, but earning a spot in the competitive career field of OSI. We shared a camaraderie that would serve us well in the months to come.

I stood in the doorway, waiting to be called in. My joke hadn't even registered with the boss.

"Come in and sit down," he said without emotion.

I took a seat directly in front of his sleek, modern desk. His computer monitor sat to his right, forcing him to turn his head awkwardly to view the screen. He scanned something on the monitor, squinted ever so slightly and, without moving a muscle, uttered, "You're headed to the big show, L.T." He said it as a matter of fact, without emotion or preamble. He had also called me by my rank – Lieutenant – adding to the formality of it all. And then his words hit me. I was going to war.

I was instantly both thrilled and a little nervous. I had begun to adapt to the environment in Kuwait, and had painstakingly fostered meaningful and sensitive relationships with high-placed sources in the government and military. I also now had work on the mosque bombing and the brewing civil war. The last thing I wanted to do was turn that over to someone else. But deployment to a warzone was the reason I had begged to deploy.

"Since one of our guys was killed in Iraq," the boss explained, cutting off my train of thought, "the regional commander needs to shuffle people around, until they can get a permanent replacement in country. You are going to fill in until that happens."

We sat in silence for a moment. I sank back in my chair. The thought of replacing a fallen comrade was sobering. Special Agent Daniel Kuhlmeier had been traveling in a two-vehicle convoy through Baghdad when a vehicle-borne IED struck his truck, killing Dan and his Army driver. A former Marine and now civilian agent with OSI, Dan was a sharp, experienced veteran who had served time in Cuba, Germany, and Africa. He had been with AFOSI for nearly three years. He left behind a wife and young daughter.

We had been at Kuwait International Airport to receive Dan's remains as they transited through to the United States. It was 3:00 a.m., pitch black save for the lights from the plane, and silent but for the low roar of the engine. Our whole office stood quietly at attention as they walked the casket off the back of the aircraft, covered with a pristine flag. Military form and decorum were foremost in these moments, regardless of your location. Appropriately, Dan was carried by Marines, an Army soldier, and three Air Force OSI agents – all three services in which he had served.

I didn't know then that I would be replacing him. This turn of events reminded me how wrenching death is for OSI. I sat in silence as I considered how the organization was a small command of top-shelf talent who had persevered through an extremely tough selection and training process. I was one of only twenty Air Force ROTC cadets taken nationwide in 2004, and only fifteen of us made it through the laborious, year-long background investigation that followed. We were multi-talented college graduates from esteemed universities. One fellow agent had originally been an All-Star pitcher in AAA baseball until a career-ending arm injury; another spoke four languages; yet another scored nearly perfect on his SAT test out of high school, garnering all kinds of attention from the media in his hometown.

I myself had served in several leadership positions before accepting the role as wing commander over one of the largest ROTC units in the Southeast, while taking twenty hours of registered classes. I went on to earn several university awards for academics, including the Distinguished Graduate Award, an honor given only to the top tier of all commissioning cadets across the country, and the coveted Air Force Cadet of the Year award presented to me by then Chief of Staff of the Air Force General John Jumper. Amid it all, I even earned a black belt in Tae Kwon Do.

And even with all those accomplishments, I was a middle-of-the-road candidate compared to those selectees that entered the OSI academy with me. Male and female, civilian and active duty, OSI took only the sharpest. So losing one was a huge blow financially and logistically, as well as emotionally.

★

"Pack your bags, you're going up against the Taliban," my boss said. "I will forward you all the details, so read them over carefully."

With that he finally made eye contact, and now I could see the big smile spread across his face. The points on his bushy mustache arched upwards, and his white teeth shone. He knew I wanted to go. He knew the regional commander had waited for the right opportunity.

I rose to leave, my head still in a fog. Erin had apparently overheard the conversation, or had known ahead of time, and stopped me in the hall. Her eyes met mine; she smiled and reached up and gave me a quick hug. "Good luck," she said and paused. "Just know that we are all jealous."

It was true. Iraq and Afghanistan were what we trained for. Life on the edge, covertly meeting with dangerous people in dangerous places, all in hopes of collecting intelligence that could save a life or change the course of the war. For those with an appetite for that kind of thing, Kuwait just didn't suffice.

CHAPTER 6

Blackout

The interior of the massive C-17 transport plane was deafening. I pulled out my Vince Flynn novel and hoped the booming engine would transition into white noise. From my vantage in the small jump seat along the wall, it looked as if the military had simply stripped an old, dilapidated passenger jet of its passenger chairs and bolted rows of decades-old seats to the middle of the aircraft. No overhead bins. No aisles. No attendants. Just rows of seating in a cavernous interior covered with wires, brackets, and metal fittings.

Some soldiers gave me sideways glances, no doubt wondering who this guy was in civilian clothes. To the untrained eye I might be a reporter, a contractor, or a spy. Furthermore, I was alone. OSI agents often traveled unaccompanied. We rarely mustered with large units or small teams. I settled into my book, and no one tried to strike up a conversation. You couldn't hear each other anyway.

At some point well into the flight, someone from the cockpit yelled to the cabin: "We're going dark!" Soon after that, all the lights went out. Pitch black. No iPods. No flashlights. Nothing. We were now over Afghanistan, and any light might give away our position in the sky and attract ground fire.

The blackout lasted some time, and there was nothing to do but anticipate an incoming missile. So I did my best to let my mind wander as I stared into the dark void. After what seemed like an hour a feeling of weightlessness overtook me, my ears began popping, and the plane made a sudden descent, dropping out of the sky as if to belly flop the runway. My stomach rose to my chest as I grabbed hold of the bar under my seat, and the familiar sensation of hitting a steep decline on a rollercoaster pulsed through my body.

I had been forewarned about this rapid free fall. Bagram Air Base sits in a bowl surrounded by beautiful mountain ranges along the Hindu Kush. The mountains themselves make it nearly impossible to achieve a slow, gradual descent, but also the enemy often sat perched in them waiting to shoot down incoming and outgoing aircraft.

The base had been around since the 1950s and had been used as a staging area for Soviet troops during their invasion of Afghanistan in the 1980s. British and U.S. Special Forces commandeered the defunct Soviet base during the U.S. invasion in 2001, and immediately used the dilapidated runway tower to call in airstrikes on enemy positions. The Americans had held Bagram ever since.

We landed without incident, and almost immediately the back end of the plane opened and workers began offloading palettes of gear and equipment. The passengers deplaned on the runway about a hundred yards from the first set of offices along the edge of the airstrip. It was chilly. The air was crisp and noticeably thin. Although it was dark, I could see the looming mountain range just beyond the base perimeter. The urban sprawl of Kuwait had been replaced by a daunting, rugged landscape full of combatants who had been in war longer than I had been alive.

We started our slow walk toward the edge of the runway, and the Special Agent in Charge of the Bagram detachment picked me out immediately. "Welcome," he said. "Let's grab your bags before everyone else gets the same idea." I spied my belongings crammed onto a massive, overflowing palette, and we snatched them up.

As he escorted me to the lodging we exchanged pleasantries, but little else. It was all so quick that I probably wouldn't have been able to pick him out of a crowd the next day. It was getting darker now, but I could make out the long wooden hut that sat on cinderblocks about three feet off the ground. He opened the creaky door to a tiny, dimly lit walkway, with rooms on both sides, a variety of faded and tattered curtains serving as doors. I peered past one curtain to see that each room was big enough for a miniature twin bed and little else.

Gingerly I followed the boss down the tight walkway, hoping not to bump one of the fragile walls with my oversized gear and send someone's belongings crashing to the floor. The rooms were sectioned off by plywood walls that didn't even reach the ceiling. It was dead silent, save for an incredible snore emanating from one of the rooms.

The boss swung wide my curtain door and gestured with his free hand for me to enter. I scanned my surroundings, which took all of three seconds. I then quietly dropped my gear. "Someone will bring you to the office tomorrow. Get some rest," he said and left.

I had not showered in two days. I used baby wipes to clean the filth from my body before lying down. I noticed a funky smell coming from the mattress. Later I learned that someone had thrown up on it, but they had no replacement, so they had simply flipped it over. The bed frame had a broken leg, so I slept on a slant. I didn't care. I was in Afghanistan.

★

I awoke the next morning to shuffling in and out of the hooch. The elevated flooring accentuated the sound of each step. The thumping of boots could be felt throughout the cabin. I dressed quickly and got moving.

As I opened the door, a cool burst of air hit me. Mountains, now clear in the sunlight, loomed large in the distance. Some of the more faraway peaks had snowcaps. I stepped off the stairs and found myself strolling along a sidewalk that paralleled a busy road just behind the hooch. I tried to take it all in. I passed old Soviet anti-aircraft equipment still anchored in place on the corner near an open-air gymnasium. Rows of dilapidated buildings dotted the roadside steps from the chow hall. Cold War-era tanks and trucks damaged in explosions decades ago littered an overgrown field. A smaller grassy expanse to my right had prominent warning signs hanging from the fence line: "Landmines, DO NOT ENTER."

It was also clear that Bagram Air Base operated like a small, congested city. Cars, trucks, and military vehicles crammed the narrow roads and the array of side streets that weaved throughout a massive expanse of compacted infrastructure. Bland-colored buildings and a mess of smaller, fenced-off areas housed special government organizations, foreign armies, and administrative groups. I worried about getting lost, so after about twenty minutes I circled back toward the hooch. No sense losing my way the first day. As I arrived back at the main door, someone tugged on my shoulder and I spun around to find a short guy with a long, unkempt beard and straight brown hair neatly placed atop his head. A smile broke out behind all the fur. "I'm Luke," he said. "You must be the guy from Kuwait."

I simply nodded.

"Follow me to the office. We'll get you set up."

We strolled through pathways dividing the maze of two-story buildings, sleeping quarters, and trailers. The path we took was disorienting, as though someone had haphazardly placed mismatched structures wherever there was space.

Luke opened the door of a hut not unlike the one I slept in, to reveal a small sitting area. And just beyond was the main cabin, an open space with large wooden desks lining the walls.

"How did you find your sleeping arrangements?" said my new boss from the back of the room. "Comfortable?" His Northeastern accent was unmistakable, although I hadn't picked up on it the night before.

He sat behind a large mahogany desk that looked as if it had once served an important person. Afghanistan had given it that weathered appearance that people paid good money for back home.

"Fine. Comfortable enough," I said, a little uncertain about his probing.

"Not the Crown Plaza." He roared with laughter. There it is, I thought to myself.

Every agent in the room turned and stared. The glaring was burning holes through me. A wary smile surfaced. I felt confident in my work there, though embarrassed about the accommodations.

Luke deescalated the situation. "Sit here," he said. "This is Frankie's desk. He was best friends with Dan; he's escorting the body back to the States."

★

"Welcome to Bagram Air Base," the Army captain said.

All new arrivals had to attend the introductory security meeting held in a small, ramshackle conference room with stadium seating to accommodate about twenty people comfortably. I was there with about ten uniformed personnel. Some of them gave me sideways glances, just like on the plane. I ignored them and tried to look uninteresting.

The captain quickly jumped into security measures, informing us of the foreign military presence and the "no go" zones around the base. Any fenced-off, wide-open pastures were littered with landmines, and some areas held sensitive people and information. He rambled on about mortar fire, bunker locations, and the alert system. Then he nodded briefly in my direction. "For those of you that play outside the wire, be aware of new Taliban tactics." Indeed, OSI essentially did all its work off the base. Most military jobs had no reason to leave the safety of the installation, but we left almost daily to collect intelligence, meet with sources, and conduct operations.

The captain continued: "Al Qaeda and the Taliban have engineered new ways of killing you by causing a traffic accident at some intersection or cross street, generally with small convoys. When you get out to check on the other driver, they send a second vehicle loaded with explosives – and BOOM." I filed that one away. "Also, know your location and potential safe havens with friendly villagers." This last statement caught my attention. Operators are always keen to know where a team can retreat in case of emergency. Stateside, that was usually near police stations or hospitals. Not here. Go to the wrong police station, and you're liable to be auctioned off to the highest bidder. So I listened intently as he described how a combat team of four guys had recently been pinned down in the middle of nowhere. Three of the four had been killed, but one survived because a village took him in – a cultural obligation called *Pashtunwali,* which dictated that they care for and protect someone in need. I was skeptical, to say the least. A village, willing to defend and potentially sacrifice its own people in a battle with Taliban forces, for the sake of one U.S. serviceman? Probably got his facts wrong.

Lone Survivor, a bestselling novel about Operation Red Wings and the survival of Marcus Luttrell, would hit bookshelves fifteen months later.

★

I had been in country for about two weeks now, sporting a thick beard and armed to the teeth. I had an M-4 long run rifle slung across my chest, a 9mm handgun in my thigh holster, and eight full mags in various pouches of my Kevlar tactical vest. In addition, we were all sporting about 60 extra pounds of gear. We had gathered in a circle just outside the office, four agents awaiting a security briefing from the boss.

"All right, fellas, you're headed to the Embassy compound in Kabul," he said. "Do your comms check, review the route again. And Dave, you drive the second Suburban."

"Roger that," I said calmly, as if I'd been there before.

There would be four agents in a two-vehicle convoy of extremely heavy black Chevy Suburbans with after-market armor retrofitted onto the outside to protect against enemy fire and roadside bombs. Each vehicle had a massive radio antenna and a heavy-duty grille guard over the front end. The weight of the added protection created greater momentum – the vehicle would propel you an additional twenty yards after you slammed on the brakes.

"Remember the latest tactics," the boss said. "If you get in a collision with a local driver, just keep moving. Get off the X, and we will send the proper authorities to work things out later."

Luke whispered a reminder to me that the size of our convoy made it a prime target. The guy I was there to replace had been killed in a two-car convoy.

"Just remember to get off the X," he said quietly.

I had already been trained in evasive driving. I knew to leave as little room as possible between us and the lead vehicle, so no cars could wedge their way into the gap and disrupt the convoy. I also knew, and had practiced, all the turning and traffic maneuvers to protect the team. Much was required of the follow vehicle.

Luke set the speed, and his partner called out what they saw along the road. Bobby and I in the second SUV would adjust accordingly.

Luke jumped into the lead car and pulled out. We weaved our way through the narrow, congested streets of Bagram and within minutes skidded to a halt near the guard shack. Shortly, we would be outside the safety of the base and on our own. Luke displayed badges to the guard, waves were exchanged, and then, without another moment to think, we had serpentined our way through the exit barriers and found ourselves speeding down the main road leading away from the base. My adrenaline ticked up as the main entrance to Bagram popped into my rearview. I instinctively reviewed all our preparation, considering the weapons checks, communications checks, medical checks.

The main drag we were now on had sustained much damage. We spent almost as much time spotting and avoiding potholes and dangerously uneven pavement as we did potential roadside bombs. The concentration of people made hiding an ambush easy, the streets lined with debris and the mud huts crammed side by side gave attackers all they needed to disguise explosives. It looked nearly impossible to avoid any pending attack. I kept as close to Luke's vehicle as possible, as we darted past donkey carts and old European sedans that choked the air with exhaust.

We maintained a high rate of speed, dodging small craters and potential attacks. The idea was to pop up in and out of an attacker's field of vision before they could react. This was incredibly important in case an attacker had knowledge of our route. Speed essentially made killing us a bit more difficult. We had to get out of the local congestion as quickly as possible. Only when we reached the open country road did I let a little space grow between us.

The drive from Bagram to Kabul was an hour of changing scenic landscapes. I had chosen a Frank Sinatra ballad for this part of the drive. The Suburban was converted for war but still arrived equipped with its factory sound system. And I, the music eclectic of the group, came armed with a package of song. Indeed, music played a remarkable role in my ability to bridge relationships with the most unlikely – and sometimes unlovely – folks, and served me well when recruiting informants of a different generation or ethnicity. For now, the crooners-style beat added a soothing effect to an already beautiful view.

We traveled along the edge of a mountain range that reached majestically up and through the clouds hovering in the high Afghan sky. Those looming mountains soon enough gave way to a rocky, dry terrain not unlike the flat desert of Kuwait. And some minutes later, that barren land became a grassy plain that stretched for miles. Huts sprang up here and there in what looked to be a farming community as we neared Kabul. We passed one or two Afghan police checkpoints. The police stayed squatting in their famous gargoyle stance and just waved as we flew by.

Our adrenaline ticked up as we neared the capital and Sinatra continued to croon. The juxtaposition of the soothing melodies against the incoming craziness of the city was oddly satisfying.

"Coming up on Kabul, get ready," Luke interrupted over the headset. I pulled closer to his rear bumper and steadied my hands on the wheel. "All right, remember to keep moving."

Kabul seemed to appear in a flash, and before I knew it we were in the thick of things. We joined the traffic on Chicken Street – the most dangerous road for American troops in Afghanistan – and met that infamous gauntlet with little more than a tap of the brake. We bobbed and weaved. Food carts and pedestrians strolled inches from the speeding traffic. We narrowly missed other vehicles in almost every swerve. We slowed at times but never stopped, pushing our way through by gently nudging smaller cars out of the way with our massive bumper. We would not idle in any one spot. "If this doesn't clear up," the radio chirped, "we will go off-roading." I smiled at the thought as I eyed the wide dirt median. We maintained high rates of speed.

After about ten minutes of high-stress driving, we came up on an eight-lane traffic circle, a massive roundabout overflowing with old foreign cars and junk trucks, all spewing exhaust and blasting their horns. I gripped the wheel hard and whipped the SUV into the madness. Cars darted back and forth. Push, push, I could hear Bobby saying quietly. Stay on Luke's bumper. Motorbikes were everywhere, weaving dangerously between vehicles, hard to see as they ducked in and out. Suddenly, one jumped out in front of me. I swung the big Suburban hard to the left, clipping the side mirror off another car, then yanked it back to the right and nailed the gas to get back up on the lead car. "That was close," Bobby said with obvious relief.

The traffic circle spat us out onto a four-lane road off the second exit. It was suddenly calm, as if we had entered another city altogether. I ever so slightly loosened my white-knuckle grip on the steering wheel.

"Nice work," said Luke. "Almost there."

The rest of the way I stayed within inches of the lead car, and we arrived safely about five minutes later at the government compound.

"Just another Sunday drive," Luke said, grinning as we jumped out.

★

The bulk of our time at the Embassy compound was spent reviewing the information our technical experts had compiled, something that pointed to a possible Taliban attack against Bagram Air

Base. As best we could tell between a source and electronic surveillance, Taliban forces planned to detonate explosives hidden on tankers, which regularly lined up at the gate to deliver gas.

Our informants had passed us active telephone numbers, so now we could monitor the attackers at will. The old "foil-or-follow" debate ensued. The source was directly involved, so we presumed no movement would be made without his knowledge. The source also controlled some of the explosives, so the operation needed him, to a degree. Yet we still didn't know who else the Taliban would assign, or the day they would choose for the attack. Breadth of intelligence grows the longer you watch a target, so we decided to follow. No interference. No takedowns. Just observation. We chose to continue watching the source in hopes he would reveal more names, more numbers, more intelligence. We had confidence that any future intelligence outweighed the risk of the attack happening without some warning.

"We got what we need. Let's move," Luke said at the end of the three-hour meeting. As we raced down the stairwell, he turned to me: "You ready to do this again?"

Hurriedly we made our way to the parking lot, a somewhat secure spot inside the perimeter, but just beyond the safety of the cement barriers that surrounded the heavily fortified Embassy compound. Bobby and another agent had pulled the short straws and remained with the vehicles during the meeting. Someone always stayed behind because unattended cars were often targets of sabotage.

Thick trees behind the vehicles gave us a measure of cover, but we still had to move fast while in the open. Loitering anywhere outside was a bad idea. We did a quick comms check at the staging area and piled into our Suburbans. Within seconds, we were out beyond the safety of the compound and back on the mad streets of Kabul.

The tall buildings left us exposed to elevated shooters, and the heavy pedestrian traffic gave bad guys the opportunity to blend in. Attack angles were plentiful as we passed side streets and corner alleys every few yards.

I began taking stock of my surroundings, as I had not done during our nerve-wracking drive in. The faded buildings and disheveled residents all looked weary. The buildings were dilapidated, the roads littered with potholes. An open sewer ran along the edge of the sidewalk, steps from where children played. An elderly man with decent clothes and a long beard stood on the stoop of what looked like the entrance to a small grocery. Tiny food stalls with hanging hooks for bread fixed to the edges of makeshift tent roofs dotted the side streets.

"Turning right at the next road, then we'll hit the main drag," a voice crackled over the radio. "Get ready, because it's going to get crazy again."

Luke hit his sirens and threw down the visor with the emergency vehicle lights. We did the same. I leaned on my horn as we rode up on the congestion ahead, and within seconds we were inches from the last car in a long line of slowing vehicles. Hearing the horns and seeing us in the background caused drivers to peel off, as if an ambulance or fire truck were bearing down on them. We slowed but never stopped, using our intimidating grille guard as an image for their rearview mirrors, tacitly but firmly urging slower drivers to make way.

Our horns and sirens blared. Our flashing police lights rapidly flickered a spellbinding white and blue. From our vantage in the second SUV, we could see car after car reacting to our noise and lights. A wide berth was beginning to form as we continued to inch our way forward at twenty to thirty miles per hour. "We're gonna take the farthest lane to the left," Luke called over the headset.

"Roger that," I said.

Luke swung into the left lane, inches from the cement wall that served as a median. It was about four feet high and looked about two feet wide, with some small shrubbery lining the middle. I could see the tops of cars headed the opposite direction just over the wall. We hugged the median as cars continued peeling off into neighboring lanes. The road ahead finally began to clear, and our small convoy picked up speed.

The road had three- and four-story buildings on both sides, so we remained at a severe disadvantage from any elevated attack. An ungodly amount of fire could rain down on us, and we wouldn't be able to react. The crowded streets and many windows also gave perfect cover for surveillance. People in the buildings or on the streets could call ahead with our location without much problem at all. I took comfort in our surveillance and counter-surveillance training. But we could be in trouble if someone knew when we had left the Embassy compound and guessed our route correctly. A good friend, a fellow OSI agent, was killed by a roadside bomb in November 2007 because someone knew the convoy's arrival time. I pondered these things briefly but shook the thoughts loose. We had to get out of this kill box as soon as possible.

We achieved a consistent speed, forty to fifty miles per hour, and the lane ahead looked clear. The heart-pumping chorus to Eminem's hit song "Lose Yourself" now flew from the speakers. I pulled within inches of the lead vehicle and settled in for a race out of this dangerous situation.

Just then, my peripheral vision spied an Afghan minivan about fifty yards ahead, moving opposite to all the other vehicles. As the wave of cars moved right, this guy was weaving through the traffic to the left and toward our lane.

"Suspicious car about fifty yards up," I called out.

"I see him," Luke replied.

We kept our speed and pushed forward. As we closed the distance, the Afghan van continued its merge left and picked up speed. We were only about twenty yards out now, and it looked as if the driver was angling directly for our convoy. He was driving one of those old Afghan vans with faded orange and yellow stripes running along its sides. The driver wore a traditional headscarf and a beard. As far as I could see, the rest of the van was empty.

"Here he comes," Luke said calmly.

All my senses were heightened. My heart began to race. I gripped my steering wheel tight, preparing for impact. I could tell by the speed and distance that this guy would try to split our convoy. And his direction made it clear that if he missed us, he would run directly into that wall. He wanted us – no doubt about it.

We increased our speed and tried to blow past him, but he had timed it perfectly. Seconds later I could see the whole of his vehicle as it passed between two cars on our right, as if time had slowed down for a second.

He missed Luke's vehicle by inches and put his nose right between us. "Hang on!" Bobby yelled.

I floored the gas.

A bone-chilling crunch shuttered through the vehicle as our massive, up-armored Suburban ate its way through the left side of the oncoming van. The collision caused the driver's head to whip like a rag doll against his side window, shattering it. Every other window in the van exploded, as if a bomb had gone off inside. Glass spewed in all directions, flying across the hood of the SUV. The impact sent the Afghan van airborne. It spun like a top as it flew through the sky, doing at least one full rotation, before landing about twenty yards behind us. I fully expected to hear and feel an explosion. But after a few moments of anticipation, nothing happened. The vehicle just landed behind us with a metal-crunching thud. We lost hardly any speed and, seconds after impact, we found ourselves right behind the lead SUV.

I glanced in my rearview and saw the wreckage, smoke billowing up from where the engine had been. A bloodied middle-aged man with a long beard stumbled from the mangled van and began weaving his way back toward the sidewalk. Bystanders leapt from their vehicles to render help. Paying them no attention, he darted out of sight.

The crash site grew smaller and smaller in our rearview mirror, and within minutes we were out of the kill box and outside Kabul.

"Heck ya bro," Luke screamed over the radio. "You earned your street cred today."

"Just doin' my job," I replied, trying to act as calm as possible, though my adrenaline was pumping. Act like you've been there.

★

"What in the world happened?!?" the boss screamed as we pulled up. The right side of the grille was severed. The bar that protected the headlight had snapped off and was bent upwards at a 90-degree angle. It looked as though some superhero had ripped off the pipe and pointed it upward to the heavens. It had taken a massive collision for that to happen, and the boss knew it.

"He did his job, boss," Luke said confidently, before I could answer.

We explained the story to our unsettled supervisor. He paced nervously as we recounted the details.

"All right, Grantham, write it up and send it to me. We may need to send somebody out there to check on this guy."

"That was intentional, boss," Luke said. "That guy was gunning for us, trust me."

Luke's words consoled me a bit. But I also knew that any repercussions were beyond our control. Even if it had been a planned attack, the Taliban would likely paint it as an innocent person forever damaged by American aggression. Traffic accidents had more than once turned into media-fueled protests. And no one wanted to be the source of an international incident.

As I began to draft the report, my thoughts got away from me. I considered how the Iraq war had sapped the Bush administration's political capital. I prayed that I had successfully maneuvered through an attack, and not severely injured an innocent man. I tried to rest, but the implications consumed my thoughts.

★

The following day, reviewing the timeline of events, I recognized that I had done exactly what I was trained to do, all within the rules of engagement. I had protected my team and finished the mission. And I would rather be judged by twelve than carried by six, I stubbornly thought to myself.

I then switched from the classified network to the unclassified one and glanced over my personal emails. The slow and often interrupted Internet connection made for a delicate situation. I was careful not to click on any email I didn't intend to read, because each message could take five minutes to load. As I scrolled I zeroed in on a message from my parents, realizing that I hadn't sent them an "all is well" email in several weeks. Probably should do that. What I read stunned me.

> Son, hope you are safe and sound. Whenever you get a moment, shoot us an email, just one word letting us know you are good. Also, your friend Jamie from high school (remember her?) contacted us out of the blue today. She said that she had a sudden, overwhelming urge to pray for you yesterday. She said she stopped what she was doing and prayed for you right there in her living room. Cool story, right! But we hope everything is ok.

The email had been sent the previous day.

With the time difference, she would have been praying for me not long before we left for the embassy. I reread the timestamp and tried to rationalize it as a coincidence. I sank back in my chair and nervously tapped the edge of the keyboard. I certainly could not explain what had happened over an unclassified channel, and I was sure not going to share a story like that with anxious family members. Leaning forward, I calmly typed a reply. "Everything is going well. Please pass along my thanks to Jamie for her prayers." I clicked send and powered down the computer.

No sooner had I risen from my chair than a massive explosion rocked our office. Instinctively I ducked, as if that would have done any good. The concussion rattled pictures loose from the walls and sent supplies tumbling to the ground. We didn't wait around to clean up or assess the damage, though. The building was still standing, and no one was wounded. So everyone in the office threw on their gear and scrambled out the door toward the vehicles. "Contact perimeter security!" yelled the boss. "Find out where that rocket came from, now!"

OSI was accountable for force protection, which made us

responsible to collect intelligence to protect the base and respond to events. For now, six agents piled into the two SUVs parked outside. The broken grille rose above the top of the hood as a reminder of what I had recently escaped. And now, we were going back into the fray.

"Perimeter security has no idea," Luke called over the headset. "No impact sight and no casualties that they're aware of."

"The warning sirens didn't go off, either," I reminded everyone.

"A landmine wouldn't have caused that kind of sound." The boss paused. "Let's go."

The boss was referring to the almost weekly landmine detonations that brought injured locals to the base hospital for lifesaving treatment. Afghanistan is a land of buried explosives, left over from the war with the Soviets. Even the busy thoroughfare near the main entrance to the base was littered with mines, despite the constant foot traffic from the nearby town. That landmines could remain in such a congested area was a testament to their abundance. Staying on paved roads or clearly marked trails was no guarantee of safety.

Passing through the front gate, Luke swung the vehicle left along a dirt road that ran parallel to the base perimeter. I swung my wheel and hit the gas, making sure to stay on his bumper. Dirt and dust spewed up as our trucks sped toward the main village on the edge of Bagram. The dusty road was strangely smoother than the paved ones. We slowed at a bend and darted right.

Before I knew it, we had pulled up on a series of large adobe huts all built at the foot of a small mountain. Other huts were dug into the incline. These villagers were invaluable as our eyes and ears for base security – the ones we could trust, anyway.

The boss and a few other agents darted for the elder's hut, with our veteran linguist in tow. Crowds were already gathering around our vehicles. They had also heard the explosion. "Where did it come from?" I heard our boss asking the elder as he approached. The conversation trailed off as the group began walking away.

I turned my attention to security as locals circled our position. Instinctively I brought my gun to the ready and scanned the crowd for suspicious-looking characters. There were two dozen people scattered about, all men, and they examined us closely. No one stood out as an immediate threat. But several had a distinct look in their eyes that drew my attention. I began to fear that the explosion was a ploy to bring us into the open, only to mow us down. The longer we loitered, the more the thought took hold.

"Tell them to stay back," I instructed our other linguist standing to my left. The longer we stood there, the more confident villagers grew about approaching us. He barked orders, and they started backing up. One man kept coming, waving his hands and saying something I couldn't understand.

"He say it come from the sky," my linguist told me. "Not a rocket, just a boom."

"Thank you," I replied. "But tell him if he comes any closer, he may get shot."

The man stopped before the translation. Apparently he understood enough English to interpret that.

I did my best to discern intent from body language and attitude. We had seconds to decide who wanted to kill us and who was just being helpful. That's a difficult position, made easier only by training, experience, and some intuition. I had already determined that he wasn't a threat. But somebody watching could very well be one. And if they judged me to be too trusting or lax, they would likely approach in the same manner, with different intentions. Be professional, but be ready to kill in a split second.

Time passed as the boss continued interviewing the village elders. Another agent and I decided then to switch positions. It was advisable to change perspectives frequently. It was easy to get comfortable with your viewpoint, your landscape, and to miss something important. New scenery helped sharpen your eyes and lengthen your attention span.

I now scanned a smattering of mud-brick homes built into the mountain face some fifty to seventy-five yards away. Numerous openings serving as doors and windows presented a serious security problem. They provided endless options for a sniper, or for multiple snipers. And close quarters meant they could quietly slide from house to house. I accepted that I would be the first to fall if a shootout occurred. Hopefully they would shoot me in the Kevlar vest covering my chest.

The homes directly in front of me also posed a problem. But I could see them and the activity around them well. Just then, a kid darted out from a door directly in front of me. I instinctively pulled my gun to half-ready position, then just as quickly relaxed. He was retrieving a red bouncy ball that had escaped through the door. We exchanged smiles as he scooped it up. He then looked at me for a second and extended his arms, offering the ball to play. But quickly he turned back inside at the sound of a voice calling for him. In clear view of the open door, his mother streaked across the room and scolded him for leaving the house. She stood five feet or so inside the doorway, barefoot, her hair down, wearing black knee-length shorts and a faded white shirt. No burka.

Our eyes met. A look of horror came over her face. I hadn't looked away fast enough. Yep, this is bad. An American male, obviously not her husband, had just seen a Muslim woman without her covering. Great, another international incident. My boss in South Dakota was going to get that phone call he had warned me about. But just before I shifted my gaze, she opened her mouth as if to say something. And then, without a word, she grinned from ear to ear, gave a little giggle, and ushered her kids into the next room, out of sight.

I let out a hearty laugh.

"What's so funny?" my linguist asked from somewhere behind me.

"I've sinned," I replied over my shoulder.

He laughed and turned in my direction. "What do you mean?"

Before I could explain, I heard the boss yell, "Load 'em up!" And I spun on my heels, tucked my rifle against my chest, and made haste for the vehicles. We jumped into our SUVs and darted back to the relative safety of the base.

"It might have been one of our jets," the boss explained over the headset as we drove back through the main gate. We later learned that a U.S. plane had indeed somehow caused a sonic boom that shook the entire base. Its pilot received a severe reprimand.

★

That evening began much like many others, with leisure time around a dartboard. I had never considered entertainment in a warzone until I arrived in Afghanistan. Kuwait City was a full functioning metroplex, with all the urban amenities. We took advantage of most leisure activities there as we would have back home. In Afghanistan we were tethered to a base, with limited opportunities.

After racking our guns and loosening our belts, we gathered in a corner of our rickety little office space. Darkness had long since descended over the base as we positioned a few chairs and the couch toward the target. "Crown Plaza have a dartboard?" Luke asked with a shaggy grin. The lighthearted jabbing began almost immediately. But serious conversation would inevitably follow. The wisdom passed down to me in those carefree moments was unforgettable.

The primitive Armed Forces Network (AFN) ran in the background as our comic relief. The Pentagon's limited programming was famous for its self-promoting advertisements that rivaled the cheesiest infomercials. We had memorized some of the most laughable lines, which became inside jokes often repeated at inappropriate times.

A sense of reprieve came with our late-evening ritual. In the intelligence world, operators are always analyzing, planning, trying to decipher the enemy's intentions, based on hard-earned and often incomplete information. Paperwork is never-ending, and sources constantly remind us that, unless we react quickly, the next threat is at our doorstep. I discovered just how jarring it can be to shift between war and calm. And, whether day or night, I learned quickly that the barrage of information demanded moments to clear our heads. Strangely enough, darts bridged that gap after the sun fell behind the mountains. It provided a momentary escape when the base went still.

The following days demanded lots of writing. One hour of fun required three hours of paperwork. And leisure or work breaks were different during the day. Lifting weights was useful, but the static repetition was not necessarily congenial to my sporting proclivities. I ventured onto the small volleyball court nearby for a game. It was carefree but not terribly enjoyable, partly because it required shoes. The sand was a sort of minced gravel that shredded the soles of your feet. You could always spot the new arrivals, because they were the ones playing barefoot.

It was during one of these down times that Luke and I invented Bagram Ball. The public affairs group often had boxes of new toys that were used to build friendlier relations with surrounding villages in their "Adopt a Village" project. We agents dubbed it "Adopt a Riot," because kids rushed the vehicles at delivery. Somehow, several cotton-stuffed baseballs from this stash found their way into our office. We spent the first few days casually throwing them around the room. Then I discovered a discarded PVC pipe, about the length and weight of a baseball bat, sitting next to an abandoned trailer. A thought came to me, and I snatched it up. Back at the office, I surveyed the landscape outside. Our small pebbled parking lot faced a two-lane gravel road that ran between our row of offices and a two-story building that resembled a sketchy motel. There was just enough space between the black Suburbans for a pitcher and a hitter. I measured off distance between the two spots, grabbed some discarded plywood sheets, and dropped them in place of batter's boxes, to avoid twisting an ankle awkwardly on the rocky terrain.

Bagram Ball was born.

The game evolved into a full-fledged, competitive home run derby. Air Force reservists from New York City, our commander, a squirrelly contractor from Chicago, a Pakistani linguist, all tried their hands at the new sport. People in the motel-dorm across the street began pulling out lawn chairs or leaning over the railing on the second story to watch. They cheered home runs, which we considered to be anything that hit their building.

Leisure activity during long missions "outside the wire" was generally inadvisable. But a few days later, on a trip back to the base, a vehicle in our large convoy suffered a flat tire. We were stopped on a road that ran beside a winding river, snaking for miles through the valley between two mountain ranges. We could see where the highway began to wrap around a rock face about a hundred yards ahead and disappeared around a bend. The group fanned out, taking security positions up the road, in the rear, and across the creek for a high position on the side of the adjacent mountain. We rotated between security and relaxing near the vehicles.

"All right, stand aside," I heard someone say off to my left. As the tire was being repaired, someone had pulled out a golf club and a handful of golf balls. Guys took turns trying to drive them into the distant hills. "Grantham, give it a shot." My errant swing nearly broke the head off the club, leaving the ball untouched on the tee. Others had more success, launching balls into the distant mountainside that loomed over the opposite bank. The balls would rise high against an inviting sun, growing smaller then disappearing against the rocky terrain. Somehow it reminded me of home, where regular life continued in our absence.

★

The momentum picked up in my work, just as my return to Kuwait became imminent. We met with multiple sources over a short period and were able to identify the location of some high-priority enemy operators. We passed those targets to the "snake eaters," or Special Forces, who were responsible for kill or capture missions. Simultaneously, our team developed critical information on several locations of weapons stockpiled in villages deep in the mountains. To keep the Taliban from converting them into roadside bombs, we traveled to those areas, bringing supplies and money to villages in exchange for the stored mortars and rockets. We would find a safe place up in the hills and destroy them. The concussive force sent shockwaves through the mountains, the explosive sound echoing in an almost endless repeat until fading off into the distance. After one trip a source warned us that the Taliban had been lying in wait, but the village preferred the supplies and money over dead Americans. Sometimes luck is as good as skill.

★

Finally, the call came.

We all waited with bated breath as our supervisor chatted with the general's staff. "Yes sir. That is correct." A long pause, then we heard the boss say, "Thank you." We knew then that our operation was approved.

The Taliban plot our analysts at the embassy had monitored for months was in its final stages, and now we were going to lead the team to intercept the bombers before they could make it to the base. No one said a word. The boss finally broke the silence. "Well, it's go time." After weeks of dry runs, covert source meetings, and surveillance, we would get to execute the plan.

About a week later, our informant gave notice that the Taliban team would be on the move the following day, and we prepped our gear, communications equipment, and weapons. Some NYPD reservists – who, after living through the September 11 attacks, were anxious to get their hands on some Al Qaeda supporters – were slated to assist us; we notified them. The boss led a quick review of our plan and told us to stand ready.

Early the next morning, we moved into position behind a massive sand berm beside the roadway leading to the base. Ten men formed a line prepared to rush the car when it came to a stop. A separate vehicle would be stalled out in the middle of the road, to force the bomber to yield long enough for us to overwhelm the car.

My heart raced as we stood there awaiting the signal.

"Target approaching," our lookout said calmly over the headset.

Seconds later, we heard it: "Go, go, go!"

We rushed the vehicle from all sides, as the panicked terrorist recited a prayer at the top of his lungs. Afghan music blared from the car speakers, heightening the drama of the moment. I quickly circled the vehicle to the driver's side, nearly slipping on the loose desert sand that made the roadway slick as ice, jerked the car door open, and yanked the man from behind the steering wheel by the collar, violently twisting him so he belly-flopped. I dug a knee into his back as I yelled in the local dialect, "*Daz a mean beft*" – Get on the ground. I had to treat the source as if he were part of the attack team.

It was over as quickly as it started.

CHAPTER 7

Gone Native

I deplaned out the rear of the cavernous transport plane shortly after it touched down in Kuwait. A little over a month operating under the Afghan sun had left my skin dark and weathered. My face was now obscured by a thick beard and a dirty ball cap. I could see my boss sitting patiently, legs crossed, engrossed in a novel, as I entered the terminal at Ali Al Salem Air Base. He looked up and scanned the cluster of newly arriving passengers, then returned to his book. Silently I approached within a few feet of him. "Hey boss," I said in a gruff tone, startling him.

"Holy crap," he said with a big grin. "I didn't even recognize you." We exchanged pleasantries, then corralled my bags and made haste for the vehicle. We had to get back to the office to return some gear and begin the laborious process of "back briefing" leadership on the mission. We preferred to do this as soon as possible, while memory was fresh.

The layout had been rearranged. A small, circular wooden table now sat where the lounge chairs had been. Four plastic fold-out seats surrounded the table in a haphazard fashion. Strange setup for a debriefing, I thought.

Before I could consider it further, our second-in-command, Rich Carven, walked in. A cheery man of some size, Rich gave me a firm handshake. "Look at that beard! You've gone native," he said, laughing his way through the comment.

We chuckled as Rich took the chair to my right. I chose the seat nearest where I stood, which happened to corner me between the table and the wall – a susceptible position for interrogation, I realized after sitting down. No one liked to be pinned in like that; it was the very reason we often used the technique in hostile interviews. But the mood was jovial; everyone seemed excited, so I didn't worry about it.

Moments later, a man I didn't recognize entered the office, spoke quietly to the boss out of earshot, then greeted me with just enough formality to give himself away as a headquarters guy.

"Welcome back, David," he said, before sitting in the chair to my left. No one called me David except my mother.

The boss sat directly in front of me. We engaged in small talk for a moment, then the boss launched into a line of questions. I recounted a few of the operations and discussed how working with sources in Afghanistan differed from Kuwait. Ten minutes of this type of back-and-forth, then a lull. The headquarters man then began steering the conversation toward a different topic. I had a feeling his presence was about more than simply an extra set of ears to hear about Afghanistan.

"Do you feel you worked your sources effectively?" he asked. It seemed a leading question without any discernible purpose.

"Well, I was working on sources developed long before I got there, but I think I was value added," I replied. The atmosphere was relaxed, but I could tell from their body language that tense discussions were coming.

A few more leading questions followed. Then we all sat in silence again.

The boss spoke first. "All right, David." Again with the David. "The Air Force needs people at Camp Bucca, and you are the one with warzone experience here in the office." Bucca was indeed the Kuwait office's area of responsibility, and we had visited the base some weeks before I had left for Afghanistan. But I had never considered that OSI wanted a permanent presence at the border base. The nameless headquarters guy explained the intention of that visit.

"Bucca is filled with Air Force units, and there are problems. Lots of problems. Big Air Force wants CI people on the ground, so they can get a better picture of the security situation in and around the base."

The Air Force was entirely unsatisfied with the information coming out of Bucca – information that directly impacted Air Force operations. He explained in some detail the litany of threats facing the airmen stationed there, and the issues plaguing their mission. "We need a better insight. We need people we can trust to relay that information." This was a high-level tasking I had not encountered before, and I was intrigued. "We also need people to tell us what the heck is going on in those cities near the base." Umm Qasr, the port town southeast of Bucca, and Safwan to the south.

An inconsequential border town made important only by its location, Safwan concerned military logisticians and war planners, since every major convoy coming out of Kuwait drove through it. "Our supply lines are getting picked off," the headquarters guy said. "Whole semi trucks are getting hijacked. Our convoys seem to get hit almost every time they drive through." His frustration altered his otherwise stoic and professional demeanor.

The boss spoke up again, in a more solemn tone. "An airman was already killed doing security for a convey."

That comment caught me off guard. I personally had run all the investigations into Air Force war casualties in our jurisdiction, which included Bucca. This was the first I had heard of an airman killed in combat in our AOR.

Turns out Airman First Class Elizabeth N. Jacobson had been killed six months earlier, several months before I arrived in Kuwait. She had been part of a tightknit security forces group deployed out of Goodfellow Air Force Base in San Angelo, Texas. She was there with a handful of others to supplement the overextended Army units at Bucca. She died after a roadside bomb detonated near her convoy, not far from the front gates of the camp. She had been providing security for the group as they made their way south through

Safwan. I was told later that the explosion had almost shaken the base off its foundation.

The Air Force was particularly concerned with Airman Jacobson's death because she was the first female airman to die in combat during Operation Iraqi Freedom, and the first security forces airman to die in combat since Vietnam.

The boss then added one more piece of information to what was clearly shaping up as my next assignment. "The major concern is the terrorist threat to the base, which may or may not be originating in the prison," he said. "We know the prison is becoming a sort of terrorist training center; we just don't know the extent of it." We were somewhat familiar with the sectarian divides within the camp, and the strict hierarchy and practices that had developed among hardened Al-Qaeda types. Those terrorists and some former Saddam henchmen held tremendous power within the TIF, and naturally the military had concerns for the base. Moreover, most people were aware that, once released from Bucca, those dedicated terrorists were rejoining their old teammates.

But it was Abu Musab al-Zarqawi, the leader of Al-Qaeda in Iraq, who had everyone talking. He had recruited a formidable band of insurgents. We didn't yet know the full extent of the role Camp Bucca played in Iraq's Al-Qaeda branch, but we knew it was significant.

The boss then steered what had now become an unusually lengthy conversation to Iran.

"We gotta know who is coming to visit, who is on base, everything," he concluded. "We especially want to know to what extent the base is compromised." He meant to what extent foreign spies might have infiltrated Bucca.

We already knew that Iran had adopted a low-level harassment policy, using local supporters to occasionally send rockets over the walls or deploy gunmen to the so-called "sniper hill" just beyond the perimeter. But Iran's main priority remained the collection of information. They took an inquisitive approach to Bucca. And they had the resources because we were in the Shia-dominated south.

We all were now keenly aware that, despite no official statements, I was moving to Bucca.

"You need to build a source network on base. Make it quick, but make it effective. We gotta know as much as we can about counter-intelligence threats to the base, security of convoys, etcetera." We essentially needed twice the profit for half the work. The typical time to groom the source, develop rapport, forge a trust, was cut in half. I had to charm them quickly, get them talking. I had to evaluate that intelligence and make it useful.

"Erin is going with you," the boss then said. I was relieved not to be alone, and genuinely pleased with their selection, since she would be the commanding officer.

Rich hadn't said a word throughout the meeting. Finally, he spoke up with one of his usual bits of wisdom. "Don't go shaving that beard."

PART TWO

CHAPTER 8

DRIFTERS

"Who are you, and why are you here again?" said the exasperated Army lieutenant in a thick Southern drawl, for the third time in our short conversation.

Erin and I had arrived at Bucca with one cardboard box of supplies, our weapons, and two laptop computers. Some OSI headquarters personnel had conducted a site visit a few months earlier, allegedly laying the groundwork for our new positions. We had been told to expect lodging and office space upon arrival. Later we learned that the visit had been little more than a six-hour stopover that yielded virtually no arrangements on our behalf.

As the lieutenant glared at us, I quietly resented the military's propensity for boasting of "overcoming obstacles" that were too often avoidable, self-induced adversity. We were paid to be problem solvers, but poor planning was preventable. Yet neither Erin nor I would dare express bitterness over not having necessary amenities ready for us. Embrace the suck, I reminded myself.

"We are investigators sent here by the Air Force," I said slowly, in a perturbed tone. Telling him we were counterintelligence agents would likely have yielded a better response. But that would also have blown any cover we hoped to maintain.

"Like I said," the lieutenant chimed in before I could continue, "no one told me you were coming. And I have no lodging available anywhere and no office space, so ..." He looked at me to finish his thought, or maybe he just expected us to accept defeat. And it was true that our advance team had left no paperwork, no orders, nothing. It wasn't uncommon in war to rely on verbal agreements and handshakes. But that wasn't working here.

I took a long breath. "My supervisor met with you, provided me your name as the main contact. I didn't randomly choose this office or find you in some Camp Bucca phone directory." Erin chuckled.

The Army lieutenant only grew more annoyed. "I don't remember any conversation about you people," he said. He then alluded to his rank for the third or fourth time since we had arrived. He obviously assumed that his title warranted a more submissive tone. And he was clearly unaware that he was speaking with two plainclothes officers.

"You will have to find room in the tent." He was referring to the massive, communal tent that sat just beyond the chow hall. The tent could fit nearly fifty people. Its generator was often inoperable, leaving night shift workers to bake during the day while they slept. I almost gave up.

"Can't do that," Erin jumped in. "We are armed federal agents. Pentagon requires that we not be grouped with unauthorized others." She was right. That instruction specifically directed that we had to be lodged with others in similar career fields: armed investigators who, by virtue of their job, had been vetted and deemed responsible to carry loaded firearms at all times, something otherwise prohibited on base. An on-base homicide two months later would confirm the wisdom of that regulation.

Erin used policy like a scalpel. She also had a knack for giving precise responses that left the listener mute. But the Army reservist had finally heard enough. His tone moved from questioning to commanding.

"Here's what you will do," he began. "You will go back to your supervisor and explain that I never met him and that we had no idea you were coming. That's what you'll do. Understood?"

Erin didn't budge. "Once again, we hail from Kuwait, sent on orders from the Air Force, the Big Air Force, and cannot simply return to our duty station. So, if you are incapable of recalling a recent conversation with our boss and refuse to help us rectify this, then please direct me to the person that can."

"Are you military!?! What's ya'll's rank?" He was hot now.

"That's classified." Erin then launched into a complicated-sounding recitation of policies used to delineate top secret clearances. She's good, I thought to myself. The lieutenant stammered, again unsure how to respond to Erin's technical policy references that prohibited him from knowing our rank.

"So here's what we will do," she continued. "Direct us to the nearest Air Force office, and we will figure this out ourselves."

He stabbed the air angrily, pointing to the adjoining trailer. "But they can't help you," he said, fuming. We left without another word.

★

War heightened the already deep gulf in expectations and mission between military branches. The Army teaches its people to take orders, to pass nearly every decision up and down a given chain of command. And that's a necessary approach for an infantry-driven service. Individuality hampers the mission.

The Air Force trains its people to do the opposite: solve problems at the lowest possible level. Such an environment breeds individuality and encourages self-direction. But it can also lend itself to ego trips and unnecessary questioning of authority. Joint operations are thus infinitely more difficult to stomach when Army leads Air Force. Conflicting perceptions of authority create problems. Erin and I agreed to avoid certain Army administrative offices for the time being.

Instead, we took to befriending civilian contractors. We found them more inclined toward efficiency and less beholden to military protocol. We went to them for practically everything: living spaces, badges, supplies. We knew how to develop sources, how to liaise with complete strangers at a moment's notice. I had been developing that skill my whole life – fourteen years of education littered with detention referrals, almost all for socializing too much. I smiled big when the OSI agent said during my first applicant interview that I would need "the gift of gab" to succeed at the job. No problem, I said.

Our skills paid off. Within a few hours of leaving the Army lieutenant's office, Erin had a place to live. Our new contractor friends in lodging management did have a bed available specifically designated for armed investigators, despite his certainty that no vacancies existed.

Erin was assigned a room with Agent Smith, an impressive and somewhat excitable Army criminal investigator, whose fiery red hair matched her personality. Six months at Bucca had only toughened her already rough edges, and we loved it. Agent Smith was as real and capable as they came.

But if Erin lucked out with a quality roommate, I hit the lodging jackpot.

During a hastily prepared meeting, one of the civilian employees pulled out a nearly pristine map of Bucca. He spread it over the top of his sturdy wooden desk set in the center of his doublewide trailer office. The edges of the map dangled over the sides of the desk like a white linen cloth atop a small breakfast table. It looked like some developer's schematic, with straight blue lines intersecting at precise points all over the page. "Males are bunked in this general area, and all I have is a spot on this row," he said, pointing to the upper left corner.

"Are you telling me that's a problem?" I said, looking at the rows of huts outlined on the map. He shrugged and explained how the last row of hooches faced the prison. My door would literally open onto the main gate, albeit several hundred yards away. I presumed he was implying that, should inmates escape toward the camp, my room would be the first line of defense. A nearly impossible feat with all the concertina wire, guards, and open space, but a possible scenario nonetheless.

"Not that it matters, but it's also the longest walk to anywhere on base," he added. "The upside – the room is empty."

I looked up from the map, fully expecting to see him laughing at his own joke. I scanned his face for some recognition. I peered dumbstruck over at the Navy officer in the room; he simply shrugged. An empty room was unheard of in a deployed environment. Only commanders or generals ever experienced such pleasures. But it was no prank. Although I had fully anticipated an unlovable bunkmate, it now appeared an empty suite awaited me. Not only that, no one could share a room with me unless they were an approved person. And this arrangement had the potential to last my entire tour.

"Where do I sign?"

The contractor just laughed and spun toward the metal footlocker against the wall. He scanned a wall of keys on the inside of the door, searching for the corresponding room number. Within seconds he located it and tossed it over the desk in my direction. I snatched it from the air with enthusiasm.

★

We secured more administrative victories over the following days. Erin managed to receive written permission to carry loaded weapons anywhere on base from the camp commander, a hardnosed Marine Reservist. Erin and I knew from experience that warzone authorizations from a base commander went further than the U.S. law that spelled out our authority to carry. A lowly airman guarding the chow hall would not question an endorsement from the commander.

Around the same time, the Air Force security forces members notified me that we could occupy a small but vacant spot in their office trailer. We all agreed not to advertise our arrangement, lest overactive base bureaucrats find out. The deal included a small corner of a short, single-wide trailer, already separated down the middle by thin sheets of plywood that reached from floor to ceiling and acted as dividers between two offices. This was not ideal for having private conversations, and it was absolutely unworkable for discussions on sensitive operations or informants. For now, though, it was at least a dedicated workspace where Erin and I could begin operations.

Most locales that managed sensitive sources had access to an established SCIF (Sensitive Compartmentalized Information Facility)

– the rooms with thick walls, top-of-the-line locking systems, alarms, and government-approved safes that hold America's weightiest national security information. We didn't have that luxury. We would have to guard our secrets in a trailer divided haphazardly by flimsy walls, alongside people not authorized to hear or read our information. We knew then that this would require creativity and ingenuity, walking the fine line between approved methods and getting the job done. Field Marshal Viscount William Slim famously wrote in *Defeat into Victory* that operating without orders or approval, while waiting for them, yet always within the overall intention of the mission, "must become second nature in war." He was referencing guerrilla warfare in Burma during World War II, but it was applicable here.

"Welcome to Iraq," I quipped to Erin as we unloaded our supplies.

★

Unpacking took little more than an hour. Relieved to have our own space, I scanned our quaint surroundings again. So far, we had a corner desk in the small base security headquarters unit, and a desk phone.

We also remained frighteningly exposed.

I expressed concern over how to function and protect our sources, an important consideration in a place like Iraq, where Al Qaeda and others regularly slaughtered informants. Indeed, tribal identities in Iraqi culture meant that the disclosure of one source could portend brutal torture for an entire group of extended relatives. At least for now, we would have to move outside the space to talk. "We need to do walk-and-talks," I told Erin.

She had been quiet, but now she raised her head. "Walk-and-talks?" She said through a curious smile. "Did you just come up with that?" She wanted to be impressed but knew better.

"Nope, *Casino* – the movie," I responded. I took pride in my knowledge of mob cinema. "They treated every room like it was bugged when the feds were on to them." Erin looked more amused than inspired.

"We *gotta* take a stroll and talk about things where no one can hear us," I said, in my best Brooklyn accent. Erin rolled her eyes.

Walk-and-talks temporarily resolved how we would conduct sensitive conversations. Digital reports were another story. That information had to be stored in a secure location and delivered back to headquarters via encrypted channels. There was no circumventing that process.

The Air Force lieutenant in our trailer had kindly let us connect our laptops to the classified lines fixed to the wall on their side of the trailer. So, theoretically, we had the tools we needed to disseminate intelligence. It was a tight squeeze, though, offering little privacy. Airmen with no office on base often stopped by to use the unclassified computer on the neighboring desk. Our only secure portal was, in effect, next to a sort of Internet café.

None of that really mattered, because the laptop we were issued proved incapable of storing the files. Lack of data space or something like that. We found ourselves forced to sketch a narrative, transfer it to a mobile storage device, then deliver it to a computer that had a classified Internet connection. We labeled a thumb drive "secret" and gave it a password. It was our secret thumb drive. Erin called it "Dave's STD."

★

"Are you guys getting squared away?" he asked one morning as he slid through the door to our humble office. Joe was a fellow OSI agent currently assigned to the prison as an interrogator; he had been on the ground for six months or so. He had worked quite hard to facilitate our transition, knowing that without some help we would fall victim to the infamous Bucca bureaucracy. If we found ourselves in a bind, Joe had the guy to call.

"We got some conversations going," I said vaguely. I explained that we had already developed some casual relationships with a handful of Iraqis working on base, and that we fully expected one of them to pay off big. He nodded: "It's a game of odds." Thus far, our contacts had offered rumors of weapons deliveries likely spearheaded by Iranian agents in Basra and anti-American propaganda operations in the port city of Umm Qasr, inspired by the Shia leader and Iranian sympathizer Muqtada Al-Sadr. But we lamented having to rely on a communal section of the office to disseminate this type of information to headquarters. Joe responded in classic fashion. "IT can get some extra drops put in." He gave us the trailer and the name of his contact. "Tell them I sent you over."

Within a few days, the on-base geek squad retrofitted our space with secret-level cable lines and hooked it up to a newly provided classified desktop. We went from no connection and data space issues, to tapping into America's classified network from the privacy of our office. The first batch of intelligence hit the wire soon after, and our bosses were thrilled.

★

Feeling empowered to circumvent bureaucracy further, Erin and I now reached out to some civilian acquaintances in the all-important base maintenance shop, to help construct and install a new wall with a locking door so we would have actual privacy. Bucca's minimalist environment meant there was little scrap wood and few tools available. Erin and I went to work befriending and ultimately bribing some workers with hard-to-come-by energy drinks.

We purposely scheduled our office renovation for evening, when the trailer would be empty. Aside from the shared space with Air Force members, there was an entire other wing of the trailer operated by a handful of Army officers who seemed intent on using our area as a thoroughfare. The plywood walls were an acceptable sacrifice, but unimpeded access to our space was simply unacceptable. We begged them not to enter unannounced. I nearly threatened to wall it off. Erin reminded me that we needed to be more strategic, since we had not received approval to be there at all. Recognizing that the more ensconced we became, the harder it would be to remove us, we decided to be methodical and quiet.

Workers entered our office long after the much larger daytime shift had retired to their rooms, and they had the job complete by 9:00 p.m.

Our Army officemates were understandably furious. Cursing and choice adjectives penetrated our wood walls the entire next morning. "We warned you," Erin responded through the plywood. "We have sensitive information that you are not privy to." That earned a loud bark about rank and other things. Al Qaeda in Iraq, Iranian forces attacking the base, Bucca prison serving as potential breeding ground for a Sunni insurgency – with all that as a backdrop, we found ourselves quarreling over transitory privileges in a singlewide trailer that would have been condemned back home.

★

In some ways, the situation mirrored the war itself. Erin and I entered a war that had become a struggle for the preservation of political capital and office space. Everyone felt the pressure. I knew even then that stockpiling political capital would lead to mismanaged priorities, and that energy would be misplaced or redirected toward especially useless procedures. And few policymakers were willing to call for a course correction. It was as if one side did not want to admit failure by changing strategy, while the other did not want to encourage change that might lead to success. Here, the problem became what went unsaid rather than what was said.

The Pentagon spent an enormous amount of energy trying to stem the tide of bad publicity coming out of Iraq. The administration issued as many press releases as possible recounting victories. But legitimate successes were many times offset by concurrent stories of civilian mass graves and the continuing slaughter of U.S. soldiers by IEDs.

"I watched a reporter go live from a supposed secured part of Baghdad," I explained to Erin during one of our evening discussions. Military operations had liberated this corner of the capital city, the defense official explained in the interview. "He was saying how schoolchildren now walk to class and vendors sell food roadside without fear of violence." The irony of it all, I argued, was that those on camera spoke of peace and security while wearing helmets and flak vests, surrounded by armed guards. "I'm not one to care much about optics, but even I would have changed that," I said.

We settled into our new office, saddled with the responsibility to aggressively collect and accurately report intelligence on a multi-faceted civil war where even the smallest military mishap could bring the Pentagon unwanted political attention. Zarqawi had capitalized on American hesitancy through continued slaughter. Iran undermined U.S. influence by manipulating the disenfranchised Shia.

I reflected on how the wreck I had caused in Afghanistan could get me court martialed here. That was the wrong instinct to have in war. But the political cowardice of policymakers forced one into second guessing a lot. I took a liking to Vice President Dick Cheney, because he seemed to fight back against this trend. Despite his shortcomings, Cheney could articulate his ideas with a clarity of judgment. His systematic dismantling of John Edwards during the vice-presidential debate in 2004 illustrated Cheney's level-headed, no-nonsense approach to confrontation. A picture of him ripped from a magazine would later hang from my office wall at Bucca.

CHAPTER 9

Tethered

It was late April and already uncomfortably warm. Several weeks had now passed since our arrival, and it was abundantly clear that the politically charged atmosphere radiating out from D.C. had infiltrated Bucca. Leadership gave us the impression that winning the war equated to not making a mess. Embodying that philosophy was a rule that essentially prohibited us and others from conducting source meets and intelligence operations off base. We hit that bureaucratic wall while setting up meetings with our contacts off base. "We apparently cannot leave Bucca without a cumbersome amount of security," Erin informed me, "and only after approval from all kinds of commanders and the QRF." The quick reaction force. The policy had been drafted with good intentions before we arrived, but leadership expressed little interest in finding a workable solution now that we were here.

"There is no way this will work," I complained to Erin. I lamented how the laborious approvals process included input from far too many people, which bogged down our planning and widened the circle of those with knowledge of our activities. Unwarranted and widespread ownership over our operations would only lead to micromanagement and, most problematic, leaks. And even if we were approved to leave the base, our sources would be greeted by a loud, rumbling American convoy. Covert meetings would be as subtle as four Humvees.

Erin conceded to the restrictions for now. "We've already wasted valuable time finding living quarters and office space promised us," she said. For now, we would remain tethered to the base.

Soon, though, a tasking came down for a convoy mission. Erin and I were to accompany sensitive material on the next trip headed to Tallil Air Base in northwest Nasiriyah, a two-and-a-half-hour drive from Bucca through the sweltering deserts of western Iraq. Although highway travel was one of the deadliest things anyone could do in Iraq, Erin and I were thrilled and quickly found ourselves at the base commissary. Like kids whose parents were preparing for a family trip across country, we stuffed chips and candy bars into our bags with unbridled enthusiasm.

On a hot April day, we loaded into a white Toyota 4Runner. I briefly considered how this no-frills truck might stand up against someone trying to disrupt our convoy. It was undeniably comfortable, with cushioned seats and adjustable steering wheel. But the vehicle had no after-market alterations. I knew it would not withstand a major collision, much less a bomb or gunfire. I also recognized that the unprotected and hardly inconspicuous SUV would surely stand out in a long convoy of drab and up-armored military vehicles. Whether terrorists felt it was an attractive target carrying important people remained to be seen.

We made our way along the serpentine route created by blast walls placed strategically along the approach to the front gate. A car could not reach full speed which, went the thinking, prevented car bombs from busting into the base. Soon after, we left the safety of the base and settled in for a long, slow drive to western Iraq. I gingerly hit the gas pedal as we inched our way forward between slow-moving military trucks. Keeping a steady pace within a sluggish convoy would be the most difficult part.

We entered the two-lane highway connecting Bucca and Tallil without fanfare. The country road was relatively straight and well maintained, nothing like the major thoroughfares of Afghanistan. I could see as we left the urban area around the base how the flat desert terrain stretched on both sides for miles. A lonely landscape, and its full field of vision allowed unobstructed views. That gave me comfort.

A light wind gently blew plumes of sand across the blacktop as we cruised at a brisk fifty miles per hour. Swirls of dirt danced gracefully above the roadway, and waves of fine sand crested then disappeared on the opposite side. A fully functioning aircon system served as our only defense against the scorching heat that now poured through the windshield. The heat gave the horizon a melted look, and I fully anticipated seeing a mirage with a palm tree casting shade over a pond. And, naturally, I controlled the music selection.

The melodramatic desert scene was disrupted about an hour into our drive. A truck had malfunctioned up ahead. "The convoy is slowing," I said to Erin as I let off the gas. The stalled transport vehicle came into view as the trucks immediately in front of us began moving to the shoulder. The wind stopped blowing and the sand quit floating, just as we came to a stop along the side of the road. Erin and I grabbed our rifles and exited our truck, taking up positions along the road.

We could see in the distance that the first part of the convoy had halted its advance several hundred yards ahead. They had only now received word of the inoperable truck. Our convoy was momentarily split. Not a good scenario. My mind returned to our missions in Afghanistan, and I wondered aloud how often one person could be part of a stalled convoy in a bad neighborhood.

It took the Humvees several minutes to circle back and establish a perimeter around the halted end of the convoy. Erin and I were instructed to return to our truck, where we sat in silence. It had been only ten or fifteen minutes, and sitting on the X made every second feel like an hour. The desert sand remained still. The smell of heated asphalt seeped in through the floorboards. The tranquility began to grate on my nerves. As the minutes ticked away, I could see that the gunners too were growing more and more anxious. "If everything is going fine, you're walking into an ambush," I thought to myself. I clung to the small comfort that an unexpected breakdown to us would also be unexpected to anyone planning an attack. Then again, remaining stationary would give an attacker time to adjust.

Twenty minutes had now passed. We could hear grumblings over the headsets about ditching the vehicles. We were asking for trouble staying here much longer. And, just as it would seem we might lose our collective cool, the maintenance guys got the truck running. Someone had put gas in the diesel engine, causing it to stall out.

★

The journey ended without incident as we pulled into Tallil Air Base late that morning. The crew of the hobbled truck would have to flush the engine clean and refill it with diesel, adding several hours to our stopover. With time to kill, we hurriedly visited the historic and immense ziggurat that loomed large over the base entrance, one of few remaining in Iraq. The massive temple built to help the ancient civilization reach the heavens stood nearly six stories high, with ascending faces on all sides, coming together to form a raised platform in the center. An intimidating staircase with hundreds of steps divided the slanted walls all the way to the top.

We decided to visit the ruins that sat within walking distance from the ziggurat, before climbing the temple steps. The walls, supposedly those of the house of Abraham, now stood only head-level at their highest point. Narrow walkways led us around a large home. I was quite skeptical, knowing that locals made a cottage industry out of duping evangelical westerners. But I felt the magnitude of the moment as we ascended the great stairs. The baked clay bricks were sturdy under foot, easy to traverse despite their age.

What I hadn't anticipated was the wind. By the time we reached the platform, I realized how the structure redirected the breeze around both sides, increasing its speed overtop. Erin and I took turns snapping photos of each other, as the wind nearly snatched my petite partner up and over the edge. The ziggurat's flat roof offered an unobstructed view for miles. I said a short prayer of thanksgiving before we descended, still unable completely to grasp where it was that I had just prayed.

We met our local OSI contacts at the base entrance and followed our escorts to their office. Envy set in as we entered their two-story, standalone building. Each agent had a sizable workspace, with a functioning computer set atop a massive desk and private quarters upstairs. Their staff did not find the amenities to be so glorious. And why should they, I laughed to myself, imagining their reaction had they seen our previous lodging in Kuwait.

Our brief exchange with fellow agents was uplifting. Erin and I had not experienced large-office camaraderie in several weeks, and we soaked up the conversations. We listened intently as other agents talked through their experiences, elaborated on enemy tactics, and hinted at ongoing operations. And now, having safely passed the material off to our counterparts, we departed in better spirits.

The short drive back to the convoy's rally point took us through an empty section of the base. As we passed through an unremarkable intersection only a short distance from our destination, a security forces vehicle flagged us down from behind. There was little infrastructure and no pedestrians nearby. Slowly I pulled to the roadside, as Erin wondered aloud what could have warranted the stop. A young and clearly nervous security forces member approached my open window and immediately began issuing a citation.

"Are you a civilian?" he asked through a shaky voice.

"No," I replied.

"Please give me your ID card."

I hesitated for a moment. The card had my real name and rank, along with other identifying information that he was not authorized to know.

"Can't do that. And why did you pull us over?"

He again demanded information. We again refused.

"We are OSI and can't provide you our IDs," Erin interjected after his third request. "And why are you writing a ticket?" she added in a tone I had not heard from her before.

He simply asked, yet again, for identification.

Finally, Erin snapped. Leaning over me with her head almost clear out the driver's side window, she announced: "You won't get anything until you explain why you are holding us up from meeting our convoy." She paused. "And don't you dare ask for his ID again until you explain yourself." The poor airman had no response to this. It was now apparent that he had been coached on a specific line of questioning and lacked the experience to ad lib.

The observing officer, whom we had just noticed lingering in the background, now approached from the rear, chest swollen with authority. But before he could say a word, Erin repeated her warning. The advisor glared at us both and, with some intensity, accused me of running a stop sign. A barrage of stunned laughter filled the vehicle. I instinctively peered into the rearview mirror. Erin also turned in her seat to scan the intersection. About twenty yards behind us sat a metal post, bent at nearly a 90-degree angle, the faded red stop sign facing skyward. "Are you kidding me?" Erin yelled. "No one could ever see that!"

He stood there, exuding stoic arrogance. I decided then that the risk of missing our convoy was enough to concede this fight. I had already deliberated and concluded that the risk of personal exposure was limited on a distant base. I handed over my ID. All eyes trained on the new officer, who nervously filled out the ticket.

I turned to Erin and said, loud enough for all to hear: "This will be my first moving violation in a warzone." I thought it humorous. Erin said nothing. I knew enough not to engage her any more. Minutes of silence followed before the trainee practically threw the pink slip through the window, clearly wanting no more of this exchange. "This must go to your commanding officer for signature," his supervisor barked, as he hovered near my window.

"My commanding officer has to sign off on this?" I said, perplexed.

"That's right. You have violated base rules." An unnecessary but wholly expected jab.

"All right," I replied.

I handed the slip to Erin. Still fuming, she leaned over me again, so the supervisor could see her clearly through the open driver's side window. "I am his commander! And I reviewed it," she barked. *Hell hath no fury*, I thought to myself as I watched her crumple the ticket into a ball and launch it over her shoulder into the back seat and return to her seat without another word. The lead officer stood there, steam coming out of his ears.

Instead of a reminder that the intersection was a two-way stop or, better yet, repairing the sign to its original position, this Air Force member had decided to use us as a training exercise in perhaps the most nonsensical situation possible. It reminded me yet again how bureaucracy and elevated feelings of self-importance were twin anchors dragging this war down.

★

Our return trip to Bucca was entirely uneventful. The wind died down just as the heat ramped up. It was early afternoon, the hottest part of the day, and nothing to break up the searing rays of the sun as it focused its relentless attention on us.

The convoy slowly wormed its way through the serpentine security barriers, before coming to a complete stop safely inside Bucca's perimeter. I saw a friend from the Air Force security team dismount a nearby Humvee just as Erin and I exited our vehicle. I sauntered over and found a miserable, completely disheveled man, drenched in sweat and nearing heat exhaustion. "What the heck happened to you?" I probed. His teammates circled around him with the same wearied look. They had taken the very unusual step of propping their rifles up against the vehicle and left the .50-caliber hitched atop the turret.

"The AC in the Humvee went out just as we left Tallil," my friend muttered. The heat trapped inside had quickly reached dangerous levels, and their body armor only increased their body temperature. "We couldn't even touch our guns the last half of the trip," he said, which explained why the guns now sat where they did. Once again, a small thing like functioning AC could have made the difference between success and failure on a mission. "Our AC was so cold that it chapped my lips," I responded. He returned my comment with a stunned smile and a shake of his head. In a warzone, there were few boundaries when it came to ribbing each other.

CHAPTER 10

SEND FLOWERS

"Mister David!" Mo darted for me, waving his hands and yelling as I approached the basketball court in Bucca Square. Mo, short for Mohammad, was a young Iraqi employed on base and one of our earliest informants. He acted as our eyes and ears in the nearby port city of Umm Qasr, where he lived, and where sources faced real risk of death from Shia militia and Iranian surrogates.

Mo was short and stocky, with wavy, untamed black hair atop his rather large head and perfectly round face. His extroverted personality made him almost uncontrollable but also gave him an unassuming, even clumsy appearance to the unsuspecting observer. People who encountered Mo often mistook him for someone desperately in need of a friend. Talkative and accident prone, he had a way of charming people. But under that façade lay a smart kid with exceptional attention to detail and a clear understanding of what we needed. The intelligence he possessed outweighed his awkward nature just enough to make the relationship worthwhile.

Mo proved useful for low-level operations and would likely never be considered for a stint in a deep cover role. But we were not in need of that yet anyway. Instead, what we desperately needed was basic information that would give us insight into local dynamics that flowed across the surface. He had access to the city, did not arouse suspicion, and followed instructions. More importantly, Mo seemed fine with the fact that our rapport was the only compensation we could offer.

Mo would also become my earliest experiment in verifying a new source in a warzone and under our compressed timeline.

The traditional source recruitment process was lengthy. Ideally, testing the reliability and veracity of an asset would be done in stages, with a handler assessing a source's access to information. Controlled tests might follow for those individuals found to have firsthand knowledge of certain information. Agents would arrange a controlled test, a scenario in which they knew the outcome, such as instructing a source to deliver a trunk full of expensive items to a specific location – say, a parking garage or public park – without their knowing the contents or the trip's purpose. The source would be under the impression that his handlers were clueless as well. Everyone involved would seem intent on simply completing the task.

In reality, those bags would contain a predetermined amount of cash or other valuables known to the agents. And the source would be monitored closely to determine his relative smarts, adaptability, and trustworthiness. If the source deviated from the route, reacted poorly to an unplanned obstacle or, worse still, absconded with some of the cash, handlers would know. In the next stage the source might be introduced into real-life situations, with specific instructions on obtaining precise information. Other sources might be present, unawares, and all of them tasked with gaining information from each other. These tests grow in scope and responsibility, until the source is let loose on major operations.

Erin and I conceded early on that we couldn't employ such a structured and lengthy process, at least not for low-level sources. We shortened the testing phase and relied heavily on intuition for the rest. It worked well with Mo.

★

"I have something for you," Mo said as we met in the middle of the court, launching into a hurried story about suspicious men he had encountered at an open-air market in Umm Qasr. With exaggerated gestures and in a high-pitched voice, he explained how some men had shown up asking questions about American soldiers' locations. With every word he grew more animated, furiously digging around in his pockets, unable to contain his enthusiasm.

I began to worry that Mo's excited energy was going to attract attention. I tried to give any onlookers the impression that our conversation was nothing more than friends having a good laugh. No one was within earshot, so I was more concerned about body language giving something away. Local workers congregated in the shade of a building about thirty yards to our left. Military personnel stood in a line on the opposite side, waiting to buy sandwiches from the little doublewide delicatessen. No one seemed to be paying us much attention. But you could never be certain.

Few on base knew our actual purpose for being there. Most, as far as I could tell, dismissed us as some sort of contractors. But eventually our regular interaction with locals would attract attention. More importantly, if our intentions became widely known, it would be difficult to secure more important, more useful sources.

To preempt these possibilities, I made sure to interact frequently with foreigners on base. Not only did I need to establish my bona fides within these groups of potential sources, I also had to make my interactions look routine to outside observers. Seeing me having lunch with an Iraqi, for instance, had to become so ordinary that no one questioned it. That strategy paid off, as I quickly earned a reputation among non-Americans as a socialite. One Army officer jokingly recommended that I look into work for the United Nations. And knowing this gave me confidence that our conversation probably looked quite normal.

The sun felt especially burdensome, as Mo continued to talk. Briefly I contemplated whether the cement could melt the soles of my boots if I stood there long enough. But I shook free of my distracted thoughts and refocused on Mo, continuing to feign calm. But my big, silly grin had little effect. Mo just plowed through, speaking fast and haltingly, jumping rapidly from thought to thought, almost forgetting to make eye contact.

I continued to smile and laugh every few seconds, for the sake of appearances. Finally he looked up, thoroughly confused. He read my face for the first time and searched my eyes for affirmation that his information mattered. I placed my hand on his shoulder, knowing the positive impact of physical contact, especially in a Middle Eastern culture where personal space is unheard of.

"Not here, remember," I said calmly and quietly. I could see in his eyes that it immediately registered. "Meet me at the empty classroom in twenty minutes."

★

Meeting locations can be the greatest challenge for intelligence operations. Spies must go to great lengths in evaluating the proper place to meet sources. And even the slightest suspicion of compromise can erase all the hard work.

We called it the schoolhouse. It was about the size of a small, old-timey school in an American farming community, where students of all ages congregated in one large classroom. It rarely attracted attention, standing alone on an empty patch of ground in an ignored corner of Camp Bucca, and a soldier or airman finding himself in this area of the base generally just sauntered by without giving it a second look.

It was a standalone structure of excessive concrete, a staple of Iraqi architecture. Weeks earlier, we had secured a key from a friendly contractor who himself was unsure of its intended use. Inside it reeked like a shed or garage left unopened for some time, and tables and metal chairs were strewn about. The concrete gave it an impersonal feeling, like a Cold War bunker, but for the faded red paint, with hints of yellow and green speckles, along the back wall. Tall ceilings and dingy windows gave it a deep, hollow appearance, enhanced by the flickering lightbulb that swung lazily overhead. It was perfect.

"Mo, remember," I told him there. "Don't act alarmed or excited when you talk with me. You gotta act nonchalant, all right?"

"Yes, Mister David," he replied with conviction. He then paused, as if my comment had suddenly registered. "What is nonchalant?"

"It just means stay calm, act cool," I explained.

"Act cold? Like the weather?"

"No. It just means stay calm and relaxed." I paused to make sure he understood. "Now, you mentioned something about Umm Qasr?"

"Yes, yes, lots of talk," Mo said. He then launched into an account that amounted to a bird's eye window into the tense atmosphere that enveloped the city. The Iranians had an overwhelming influence there. Some people were claiming Iranian soldiers frequented Umm Qasr, although no one could be certain, since the alleged soldiers always left as quickly as they came. Residents assumed they were Iranian, because they spoke Arabic with an accent and looked military in their gait and appearance.

Mo went on: "Different people are intimidating locals, telling us not to work with Americans." Shia militia roamed the streets, leveling veiled threats of reprisal against any who chose to work with Americans or even be employed at Bucca. The intimidation sounded infrequent and uncoordinated, though. It wasn't terribly concerning, since that happened everywhere in Iraq. From Mo's stories I discerned that the whole thing was a message not to get cozy with U.S. forces just because they were down the road. These goon squads tried to intimidate residents into believing that Umm Qasr remained Shia territory under the thumb of Iran, and that the Americans operated nearby only because they allowed it.

That was not completely off base. Between our intelligence on Iranian activities and what we knew about American military policies, Iranians were using their local surrogate militias in the south quietly to exploit the waning of the Pentagon's appetite for conflict. Weapons smuggling, IED manufacturing, kidnappings, and other covert operations grew, in place of direct, sustained combat with American forces. The Islamic Republic was more interested in intelligence collection and indirect provocation. An IED built by Iran, but put in place by a local terrorist, gave the mullahs in Tehran some deniability.

That strategy seemed to be working because of the particularly thorny geopolitical issues that beset any U.S.-Iranian conflict. No one wanted a broader, regional war, and no commander in Iraq wanted to be responsible for sparking one. The reactive mindset put us in a bind because intelligence is inherently proactive, a subtle form of provocation. And the current atmosphere in the south was not conducive to much preemption, which to us felt like surrender.

So something akin to wartime shadow boxing ensued. The British had already achieved something similar in Basra, about one hour to the north, probably helped along by more and more expressions of regret for having backed the invasion from the British hierarchy.

I was convinced that Iran intended a long war, a plan to chip away at American resolve through low-intensity conflict that would lead to an early exit. Then, with Shias in power, Iran could rather easily co-opt the entire Iraqi government.

Mo concluded his rundown of Umm Qasr and confirmed what I had suspected, substantiating information from other sources.

I switched gears. "You were searching for something in your pocket back at the basketball court," I asked. "What do you have?"

"Yes, yes, Mister David," he said excitedly, plunging his right hand into his jeans pocket. I was always amazed at how locals wore long, heavy jeans, even at the height of summer.

He pulled out a crumpled newspaper and began furiously unfolding it, smoothing it out across his thigh, then held it up like a tourist looking over a map. The Arabic font was incredibly small. The words were compressed into several side-by-side columns, all surrounding a black-and-white photograph of a cleric poised intensely in front of a microphone, mouth open and hand raised.

"Those men were handing out papers to people at the market," Mo explained. "Al-Sadr telling everyone not to work with Americans."

Muqtada Al-Sadr was the foremost Shia leader in Iraq and militia leader for the Mahdi Army. His word carried tremendous weight, and any statement of his was understood to be virtually a command. What appeared to be a suggestion for Shia Iraqis not to work with

U.S. forces was, in fact, an unofficial declaration that consequences awaited those who failed to heed his instruction.

This looked to be part of that ongoing intimidation game. And printed material suggested it was more intentional than I had previously thought. I took a mental note.

I then noticed that Mo's hands were shaking, ever so slightly, as he fumbled with the paper. His voice cracked just enough. "See here, Mister David. This is bad American stuff," he said, jabbing a wobbly finger at a couple lines in the newspaper.

I gently prodded him. "Were you threatened, Mo?"

Fear and uncertainty spread over his face, as he recounted the comments made by the men circulating pamphlets in the market.

"I'm just scared for my family," he explained softly. "Many people know I work here. I just want Iraq to be free. We are Shia, and our tribe is well respected. And people are acting crazy right now."

I tried to console him, but I had little to offer. I had no access to money, gifts, or protection. But something about his speech also seemed manufactured. Not insincere necessarily, but parts of it felt concocted.

"You don't have to do this, Mo. You can walk away," I said calmly, with a look of detachment.

It was a response he had not anticipated, and one I leveled purposely. Instead of appearing overly sympathetic, I chose to be polite and curt, in hopes of gauging his reaction. If he believed his information was merely helpful, rather than crucial, he might reconsider our relationship. Or he might decide to redouble his efforts because he believed in the cause. Instead, he offered a rather unusual reaction. He looked confused and began to fumble over his words. "No, no. I want to work or do this for, hmmm, Iraq, I mean America." And his body language spoke volumes. Between intuition and training, I could tell at that moment that Mo had been talking to a lot of people.

Mo had been coached or, at minimum, encouraged to take advantage of any relationships he had with Americans. Someone had threatened him or persuaded him to use sorrowful stories, or he himself had decided on that approach. It was a good choice, since we instinctively want to believe those for whom we have sympathy. But I didn't bite, and he wasn't sure how to react. I surmised then that he was probably low-hanging fruit for someone else, much as he was for me.

I smiled and rested my hand on his shoulder. "Mo, we will make Iraq safe and I will return as a tourist, not a soldier." I was now manipulating his sympathies. I gave him something to cling to as an Iraqi, but nothing to work with as a source. Hopefully he would now go back to whoever was coaching him, unwilling to offer them any more on me.

Sadly, I never saw Mo again.

★

The exchange with Mo, although brief and largely inconsequential, made me recall a scene from the 2001 hit film *Spy Game*. The movie concerns a relationship between master spy Nathan Muir, played by Robert Redford, and his naïve understudy Tom Bishop, played by Brad Pitt. During a teachable moment, the weathered CIA chief instructs his young operative to leave the informant he had promised to extract from behind enemy lines. Bishop pushes back on the command, demanding an explanation. But Muir refuses to provide answers and forcefully repeats his orders. Bishop ultimately ditches the source, but Muir confronts him over his insubordination, reminding him to obey instructions. Clearly upset over going back on a promise to his source, Bishop responds as any of us might: "You don't just trade these people like they're baseball cards! It's not a … game!"

"Oh, yes it is," Muir retorts. "It's exactly what it is. And it's no kid's game either. This is a whole other game. And it's serious and it's dangerous. And it's not one you want to lose."

As it turned out, Bishop's source had recently begun talking to hostile intelligence agencies. Muir had neither the time nor a reason to tell Bishop before the operation. And Bishop had developed a misplaced trust in his source, a faith in the reliability of the informant over his own mentor.

Bishop is told never to risk his life for an asset. "If it comes down to you or them," Muir says, "send flowers."

Mo knew the risks. So did I. I had determined that Mo's loyalty rested ultimately with his family and his tribe. And there was no doubt that, if ever forced to choose between me and his tribe, Mo would send flowers.

CHAPTER 11

Exposed Springs

"Man, this can get complicated," I said to no one in particular. Erin remained intent on her report to headquarters.

The work had accelerated for us since establishing our base office. A rudimentary source network and aggressive collection techniques had already yielded quite a bit of information. Any good intelligence depends on multiple streams of information, and those few we had established generated an inflow of data that opened our eyes to the complexities of this war.

It was now early May, and the sectarian violence had reached fever pitch. And the confusion seemed to increase along with the violence. A short time at Bucca had taught us to pay close attention to associations and intent. Distinctions mattered significantly to how we evaluated incoming information, how we judged their aims, whether it be a Shia militiaman or a for-hire bomb maker, a Saddam loyalist or a former government official turned Al Qaeda operative. The multiplicity of actors along with their understood – and misunderstood – intentions demanded a delicate treatment of identities and information. It took some training and experience to distinguish among them, which could make the difference between stopping an attack and missing it altogether.

That was well and good, so long as we had plenty of time. But we needed intelligence fast, without sacrificing that academic understanding of all the parties involved. Distinctly alike in motives, yet completely dissimilar in execution, the Shia resentment and Sunni power made it feel as if we juggled two wars at once. The intricacies of it all created both contradictions and clarity. It pushed us to achieve a meticulous rendering of on-ground conditions.

We took to Bucca Square at night to talk up as many locals working on base as possible. Crowds of military and foreign workers would gather most evenings at the wooden picnic tables spread out along the perimeter of the basketball court and dine on refreshments and hookah from the small white trailer that served as the square's food truck.

Gathering intelligence and spotting sources in this environment required unassuming tactics and lots of patience. We rarely probed for information, instead allowing discussions of religion, soccer, or family to dominate the conversations. To build rapport, I often tried my hand at Arabic, with help from the linguist who accompanied me virtually every night. I was trying to build camaraderie with those who might know my actual job. If they came to like or respect me, they might become willing to provide information. When I felt the conversation had a chance to veer in the direction of intelligence, I ensured that my questions and the conversation would not alarm anyone. I would employ leading questions or generally suspicious inquiries that wouldn't create an uncomfortable group environment or dissuade anyone from approaching me privately in the future.

This routine went on every night for the first two months. I would escape to my trailer at seven or eight in the evening to catch a brief nap, then head to Bucca Square, often not returning to my hooch until two or three in the morning. Out of bed early the next morning and on to the office to record my conversations, report on potential sources and cross-pollinate intelligence with the others in the office, and at it again by evening.

Those initial struggles yielded a better understanding, thanks to late-night bull sessions with the Iraqis. But the new clarity made it readily apparent that, to protect the base, and ironically to protect those hardened Sunnis providing useful intelligence in the prison, we had to get off the base.

★

"Security forces said we can tag along," Erin advised one morning. We had decided that cold-approach tactics were better than no-approach. We used tagging along with security forces units doing high-risk traffic stops along the roadways just outside the gate as a pretext to leave the safety of Bucca. Shia assassination squads were a major part of the violence in the tit-for-tat civil war that had engulfed the country. Intelligence reporting was clear that teams of killers traveled through all parts of Iraq, murdering and maiming at will. Some of the operations appeared to be strategic, taking out important members of local tribes or government organizations. Others looked like just opportunistic reprisals.

The squads were usually made up of three or four men carrying handguns and driving what southern Iraqis called "the duck," a '90s-model Toyota Crown. With that in mind, our security forces would randomly stop vehicles traveling with one or more young males for a brief search.

Erin and I piled into the Hummer ready for battle. The wind blew abnormally strong that day. It proved difficult at times to hear my linguist, but we were able to establish a routine rather quickly. During each traffic stop we took the opportunity to casually chat up the driver and any passengers who stood patiently near the car. What would seem like an ordinary conversation was actually full of leading questions intended to uncover their affiliations, feelings toward the U.S. military, and other insight. If their body language or answers piqued our interest, we probed a little more and gave security forces a subtle signal to prolong their search. A casual request from us to swing back by the base if they wanted to provide information concluded our encounters.

I worried we had scared off some potential sources. More than once, my interpreter had to quell fears that I was not a Syrian mercenary. Abdul was at his best in these moments, a real salesman, waving off the armed man with the light skin and dark beard as nothing more than American muscle. But I remained optimistic. "Luck favors a prepared mind," I told him.

★

Remaining tethered to base kept us seeking new opportunities beyond traffic stops. Somewhere Erin was made aware of the scores of people that came to the prison visitor center. Crowds gathered on designated days to visit those interned at Camp Bucca, based on Geneva Conventions rules. Mingling with large concentrations of people near the base could be a potential gold mine and not cause a problem with leadership. Hardliners on the inside likely used visitation opportunities to deliver instructions and information covertly to counterparts outside the walls. The possibility, even if remote, of identifying, monitoring, and even disrupting a line of communication proved intoxicating.

I strode slowly between the long lines of people that extended about fifty yards beyond the entrance to the visitor center, an M-4 rifle slung prominently across my chest, a sig Sauer 9mm handgun strapped to my thigh, and a wealth of armor and ammo covering my upper body. I was not dressed for casual conversation, but the gift of gab, civilian clothing, and a smile helped overcome the intimidating accessories.

"So you drove all the way from Baghdad," I asked an older couple near the front of the visitor line. "Isn't that around a six-hour drive?" The gentleman wore loose-fitting tan slacks and a tired, old brown polo shirt with weathered sandals. His wife wore equally dated clothing, dark brown from head to toe, with an angelic white headscarf that seemed to beam when the sun bounced off it. Both possessed spunk that came through in their body language, and their eyes radiated an energy that seemed out of sync with their tired look.

Our interpreter relayed my question about their drive, and they both nodded profusely. They seemed very friendly and conversational, unfazed by my guns and beard. The husband commented that their vehicle had an inoperable air conditioner and only remnants of two front seats, exposed springs covered by blankets where a padded seat had been. He conveyed all of these rather startling details with a shrug. I cringed, thinking about this couple's long drive through a warzone in the scorching heat on exposed springs. I was paid to embrace the suck; they were living it. As if reading my mind, though, the man waved away any concern. "We do this all the time," he said. "Our son has been here for months."

As best I could tell, their teenage son had been scooped up in one of those famous U.S. military sweeps that occurred after any IED attack. "It's like trying to kill a mosquito with a hammer," I quipped quietly to Erin. They believed he would be released soon; his mother cried gently as she explained how terrorists were ultimately responsible. "If there were no attacks, then he would not have been arrested," she said. I squeezed the father's shoulder and offered a sympathetic glance. They seemed genuine, and our continued discussion indicated that they had little to offer as sources. Erin and I pressed on.

I returned to scanning the crowd for anyone who might fit a profile of combatant or who displayed suspicious behavior: distinctive clothing, a curious expression, eyes suggesting a sinister soul, gait, physique – anything that said they might not be regular folks visiting friends or family. We conversed casually, hoping to charm those we spoke to and, especially, those who watched. Secondhand rapport, as I thought of it.

We meandered through the lines, appealing to the crowd with disarming looks and words of comfort, while my mind stayed battle ready. The smell of body odor thickened as we moved deeper into the crowds. A dab of gun lubricant just under the nose prior to leaving the office helped dull the effect. And the day couldn't have been hotter. The sun seared the skin on my exposed arms, I gripped the rifle tight against my chest, my hands draped over the weapon in a ready position.

We decided to split up momentarily. Erin took her linguist and a newly arrived OSI agent to another part of the line. My linguist and I continued forward.

I noticed a middle-aged woman dressed in her traditional, full-length burka, accompanied by two young boys and a preteen girl. Perhaps a mother, probably too young to be a grandmother. She refused to make eye contact but made no attempt to hide her glare, annoyance oozing from her eyes as we approached. She stood rigid, bothered by my presence but unwilling to move me along. I squatted, bringing myself eye level with the two boys and smiling at the older one, who looked seven or eight. "What is your name?" He had a thick head of black hair, shabby brown clothing, and piercing blue eyes suggesting a sweet spirit.

"My name is Muhammad," he said sheepishly.

"I'm Dave. Where are you from?"

He named a town neither I nor the linguist had heard of.

"Do you like soccer?"

He nodded vigorously, before taking a moment to look up at his mother for reassurance.

"Maybe one day we can play soccer together, Muhammad."

He lit up at the suggestion, his eyes swelling with delight, but quickly repressed his excitement, as if he had crossed an invisible line. His mother refused to look down even as we talked – but, curiously, she didn't pull him away either. Finally, I raised my hand for a high five, and he responded enthusiastically. But again, he quickly retreated behind her leg, almost embarrassed at his eagerness. Finally I rubbed his head and lumbered into standing position. I tried once more to make eye contact with the woman, but she stiffened and refused to acknowledge me.

We continued to move among the crowd, casual but deliberate. I assessed and dismissed each person in turn. You must be able to scan, judge, decide in a second or two, and move on. As we neared the end of the line, I peered off in the distance at the cars idling alongside the road. Automobiles laden with explosives were all too common throughout the warzone, so everyone remained vigilant for unmanned cars conveniently parked on the shoulder or approaching vehicles that rode low. Nothing looked out of the ordinary to me.

But just before I spun around to begin walking back to the front of the line, I spied two men seated in a dark blue sedan, scanning the crowd through a dusty windshield from behind the wheel of a "duck" car like those used by Shia hit squads.

Suddenly the passenger swung his door wide and stepped out. He stood calmly behind the door, one foot still pressed against the floorboard and the other planted firmly on the ground. He looked young and fit, a little black stubble running across the bottom half of his face. Our eyes locked. He made no sudden moves, only glared through narrowed eyes. I gripped my rifle tighter, just as his gaze seemed to grow dark. I knew this guy was trouble. But maybe the heat was playing with my imagination.

I tilted my head ever so slightly and brought my rifle closer to my chest, gripping the front handle and placing my hand prominently across the trigger guard, making sure he saw it. My body language was as clear as it could be. He seemed to grip the top of the open car door in a controlled, white-knuckled rage. He then, of all things, gave me what looked like a smile. It lasted only a second. A gust of wind carried sand across the road between us.

The driver, a middle-aged man with a round, clean-shaven face and crisp white dress shirt, reached over and tugged on the younger man's pants leg. I had not paid much mind to the driver. He looked like one of Saddam's potbellied henchmen, whose egotistical waddle followed the dictator everywhere.

The driver calmly signaled for his partner to sit down so they could depart. The younger man obliged and, within seconds, the driver put the car into a U-turn and sped away. Dust lingered where the vehicle had been.

The entire encounter lasted no more than a minute. It went so quickly, in fact, that I began to wonder if I had imagined it. My heart racing, I removed my gloves and found my hands moist with nervous sweat. I spun around quickly to begin the walk back and nearly collided with my linguist.

"We have to go," he told me. "Erin found some people with information."

"Dave, we need to talk with some people," Erin told me as she approached. "A couple of guys are hinting about Shia militia meeting with some interesting people at a local factory somewhere near the border. We need to find a location to talk, and fast. They're about to piss their pants talking to us out here in public."

We gathered ourselves and bolted for our vehicles. One of our linguists, who had lingered in the shadows during our march through the visitors' line, stayed behind to deliver the meeting location to the two men. We would circle back later for him.

★

We had to meet off Camp Bucca, and our restrictions again hampered our work. Neither individual had access to the base, nor would they want to be seen entering it anyway. But the potential information compelled us to bend the rules. We quickly settled on a fairly secure spot frequented by coalition forces, near the British security zone in Umm Qasr. The row of buildings didn't technically require the approval of Bucca's bureaucracy, and we didn't seek it.

We were directed to a small building off the beaten path, unlikely to be under surveillance from those watching the movement of coalition forces. We communicated our spot to our linguist and cleared the scene. It was a no-frills, makeshift meeting room with low ceilings and dated red carpet. The colorful walls lightened up the tight space that contained only a large fold-out table and a pair of bookshelves at each end of the room. It reminded me of a Sunday school classroom at a small-town church, cozy but in need of upkeep.

The group took up positions inside and outside the building. Five in all, we checked weapons and rechecked radios to confirm a clear and constant communication back to base QRF, in anticipation that the sources would be arriving soon.

No one said a word as we waited.

There's something about waiting, when a body at rest causes the mind to stir. There was no way to avoid it in those especially stressful moments. Maybe it's more common to intelligence operators, trained constantly to consider angles and out-think adversaries. I prepared my mind for the possibility that the visitors might be attackers. But they would be searched thoroughly multiple times before reaching us, I told myself. I considered the potential information and what it might mean for us, for the war. That mind cycle would loop back on itself, as contemplation gave way to analysis of the possible intelligence, then to second guessing, then again to hopeful thoughts.

"They're here," one of the guys posted outside yelled through the door. My mind shook loose and narrowed in on the task at hand, now mission oriented, like adjusting the scope on a rifle.

Within minutes, the two men emerged at the threshold with our linguist. Erin and I remained seated across the table as they received one more thorough pat-down. They stood expressionless, unfazed by the security precautions. I offered an apologetic smile, as if to communicate my disappointment in the whole humiliating process. They returned looks of understanding. Their body language gave no hints of stress, no sweating, no five-mile stare common to bombers before detonation. My brief assessment allowed me to relax just a bit. But at this distance, we'd still be dead.

They entered, and our linguist shut the door behind them. They took seats across from us with their backs to the door. I immediately regretted our positions; they were between us and our only exit. But I didn't want to make things uncomfortable by requesting we switch. With a guard in the corner and one at the door, we accepted the risk and proceeded.

A tall, skinny Iraqi sat before us, his eyes set deep in a thin face with a prominent nose that suggested Roman heritage. His dark complexion and frail frame seemed unhealthy and jarred with his full head of shaggy, black hair. I could see he had a constant fidget, either extremely nervous or going through drug withdrawal. He spent the first few minutes sporadically tapping the inside of his arm, like a heroine user in need of a fix.

The second gentleman sat stoically, either unmoved by the meeting's gravity or completely unaware of it. I couldn't tell yet. He was rotund and short, with dark stubble and a full head of salt-and-pepper hair. Even seated, one could sense his strength and maturity. He was someone who demanded respect, someone of relative importance.

The first man launched into a scattered retelling of details. "Two men, perhaps three, not from our village, visited for a few hours, a few days ago – maybe three days – with nice clothes, but mean. Not to me, just angry-looking." His story jumped around; he offered no real sense of time or space; his train of thought was unorganized. The information spilled from his mouth so quickly that it became hard to piece it all together. I worried that this would turn out to be a waste of time or, worse, a setup.

This went on for several minutes. The conversation became more Arabic than English, as our linguist repeatedly asked him to slow down and restate something. Erin or I occasionally interrupted his scattershot storytelling as well. He had no real answers, and that concerned me. I asked again for clearer details, how he knew this information, while trying to keep a friendly look about me. But my frustration soon became visible. Had to be a setup, I thought. But we gave it one more try.

With a big smile, I waved my hand to cut him off. "You haven't answered many of our questions," I said. "So far, it's hard to make sense of your story." I was firm, because my instinct said this guy was a fraud, but I tried to convey a non-accusatory demeanor. He paused a moment after my words were translated, fidgeted with the buttons in the center of his brown shirt, then leaned back in his chair and raised his hands in a non-verbal cue of surrender – an awkward move that caught the whole room off guard.

The older gentleman, seated to his right, had remained quiet, displeasure with his partner's performance plastered across his face. We turned our attention to him. This seemed very curious indeed.

He looked at his companion and essentially told him that he had failed. Apparently the younger man had wanted to be the one to deliver the information, and his partner had allowed it.

The older man proceeded to make sense of what the first man had tried to say. Some suspicious characters from out of town were meeting regularly with known Shia militia at a school in his village. He was convinced the visitors were Iranian, based on their dress and facial features. The respect they commanded among the gun-toting Shia militia seemed to confirm this. These surreptitious gatherings had been infrequent at first, but over the last few months both men had noticed them becoming more regular.

He then transitioned into hushed tones to say that at the last meeting, the Iranians had delivered "stuff" in the beds of pickup trucks. He didn't know the contents and had watched from afar, unwilling to get close lest it cause him and his family trouble. I found the need he felt to whisper in a safe place like this demoralizing. But it was a learned habit I had seen before from my extended family in Cuba.

As he spoke, I processed the information. Firsthand knowledge suggested a level of truthfulness. The details sounded plausible. The fact that others had reported similar stories gave his narrative credence. And he never contradicted himself, even after we posed questions in an effort to make him do so. His willingness to admit when he didn't have an answer gave him further credibility.

"Why are you telling us this?" Before going further, we had to get a sense of his motivation. His shoulders slumped just a bit, a nonverbal cue that suggested he saw himself as carrying a burden. "I just want my village to be safe," he said. "And these guys want nothing but trouble. They will cause problems and then run back to their side of the border, and we are left to deal with the consequences." He seemed to care more about Iranian instigation than about supporting U.S. efforts, but that was still a motivation we could work with. "They will know I spoke with you. Somehow they will find out," he said, resigned.

Now came the pitch. He clearly had access to information we

wanted, and he was reasonably motivated to continue providing it. But we had nothing to offer. Travel restrictions and no money put us in a weak position. So I put on my best salesman routine. "Would you be willing to call us when they come back?" I went on to frame the request as mere observation. Nothing interactive. No direct contact with the Iranians. "Just be eyes and ears," I said. "To help protect your village."

That seemed to do the trick. The risk was minimal compared to the possible reward. Or that's the impression I got from him. We agreed that he would contact our linguist by cell phone, should he learn of any pending meetings or find out the plans discussed at the previous one. We would develop a code language so he could offer information without compromising himself.

How a meeting ends can often indicate future success, and a linguist's rapport can be almost as important as the operator's. We left this meeting in good spirits. We made the man feel valuable; we told him he was doing the right thing, reaffirming his suspicions of Iran. A blend of manipulation and honesty. I felt encouraged that we had found someone to give us information for nothing.

★

By the time we returned to an empty office, dusk had settled over the base. I spent the next two hours drafting a report about the day's activities. The silence in the office helped my concentration but I couldn't completely focus, because it felt so unusual. The amount of noise that surrounds oneself in a deployed environment is hard to fathom until everything goes quiet. It was at once peaceful and eerie.

I was able to wrap up my work just as night fell over the base. And, feeling inspired, I made plans to spend my evening in Bucca Square.

CHAPTER 12

Furniture Wars

"I'm not complaining," one guy told another during a casual conversation at a table nearby. I had just arrived at Bucca Square and, from the gist of the conversation that had now trailed off, I gathered that both men appreciated that life at Bucca had those disagreeable qualities common to deployment, but it was nothing like Vietnam, and much better now than those earlier days in the Iraq War.

Alone at my table, I laughed to myself.

The late-evening breeze picked up and sent a gust sweeping across the court, like a blow dryer aimed at your face. The 120-degree daytime heat had dropped to 90. Those toasty Bucca nights played tricks on you. The darkened sky gave the illusion that the heat would dissipate. But as it lingered it left you cranky, argumentative and, oddly enough, giddy. The threat of attack, coupled with day-to-day, manageable unpleasantries, chipped away at your patience, stirring up a mood that left you numb to the surroundings, yet steeped in them.

The air tonight seemed especially still. The crowd was thin, the square unusually empty. Perhaps the absence of chatter accentuated a calm that always existed in those moments of quiet.

It was nearly midnight, and I was filthy. I considered leaving to shower, with plans to return, hoping that maybe by this hour the water would have cooled. Bathing was the only routine that provided welcome privacy on an otherwise busy, crammed base. But it was little comfort knowing that the sun had baked our sole source of water as it sat in large tanks near the trailer. The water always bordered on scalding during the midday hours and remained hot through the late evening. Cold or lukewarm showers did not exist.

As I made my way to those showers, I took stock of all the satellite dishes adorning the roofs of trailers throughout the compound. I hadn't purchased one. I had no interest in doing something I could do stateside. But the reprieve it offered was completely understandable. Hundreds of television stations from around the world beamed directly into makeshift living quarters. The signal might wane, and the channels might not provide the content a viewer wanted. But it offered a return to the familiar. I felt an awkward sensation as I passed a room whose television faintly sent out familiar sounds from behind a cracked door. Somewhere back home, life went on.

I found myself more exhausted than usual as I arrived at the showers and slowly undressed in the empty trailer. I contemplated not returning to the square. The makeshift shower stalls were just wide enough for me to slide in sideways. I turned the water on and closed my eyes. The shower head jutted out only inches from my face. Tonight I had naively hoped to find relief from the searing heat, but it was not to be.

Somewhat refreshed and already drenched again in sweat, I gathered my belongings for the walk back to my room. The pebbled path crunching and shifting underfoot was the only sound I heard on this eerily quiet night. Nearing my room, I could see the lights beaming in the distance, causing the prison to glow against the dark sky.

I entered the room and resolved not to return to the square. I was early to bed tonight. I reached for Frank Peretti's book *This Present Darkness*, an imaginative take on the world of spiritual warfare. The gripping fictitious story proved so intense, and so situationally relatable, that I had to read it in small bunches. I turned to my collection of *Seinfeld* episodes to calm the nerves. The comedy put me at ease and quickly drew me into slumber.

About 2:00 a.m. I was rudely reminded of one of the other nuances of deployed life: the luxury that came with an en suite bathroom. I tossed and turned, hoping the urge would pass. Nighttime trips to the restroom required shoes and clothing. But biology always won out, so I got dressed and marched the several hundred yards to the nearest toilet.

I realized early on that I was not the only one frustrated by the long march to the bathroom. Only a few days later a detailed email was sent out, demanding that men stop urinating in empty water bottles and leaving them in their hooch.

★

The next morning I felt unusually refreshed. "Some guy just came in and asked for Cupcake?" Erin's grin stretched ear to ear as I entered the office. I stopped cold at the mention of my call sign. Her eyes sparkled with mischief. She knew that if my nickname acquired in South Dakota got out, the haranguing would be merciless. How she had come by this information was beyond me, but she was clearly enjoying herself.

"Glenn, the security forces sergeant from Ellsworth, came looking for you," Erin began, explaining that a colleague from South Dakota had come to the office and had apparently forgotten my real name. There were only a few military police back home who were on a first-name basis with me. Glenn and I had crossed paths during a few cases, so we were familiar with each other, though not particularly close.

I still had said nothing and took a seat behind my desk.

"He came in here asking for Cupcake, and I didn't have a clue who he was talking about." Erin had assumed it was another officemate, one of the portlier Air Force guys. Then she named another office worker, thinking he was referring to the weakling of the group. "I went through the whole office, insulting everyone who I thought might resemble that call sign," she said with feigned annoyance. "Finally, I said 'Dave Grantham'?"

"Well, what did he want?" I said, trying to move on.

"Oh, yes. He said he left some furniture at your hooch. He's headed back to the States and left you a bookshelf."

I shot up from my chair at the mention of free furniture. "How long ago was he here?" I barked, now annoyed.

"Oh, about ten minutes ago." Erin didn't even look up from her computer screen. As I made haste for the door, she called after me: "You'd better hurry, Cupcake."

The strangest of inconveniences in Bucca life involved the quest for cheap furniture or the scarce materials to make them. Everyone wanted a small TV stand for their room, or a tiny desk, or some sort of cheap storage box for rooms that came with only a small bed and a metal footlocker. With gear stacked high, most people just wanted something to help stay organized. The rooms offered truly minimal living space, so every square inch had to be used with relentless precision. The key was to get furniture from someone redeploying back stateside. Those lucky folks would leave furniture outside the doorway to their hooch, as a way of announcing their departure. Anything set to the side of a door was considered fair game. We routinely circled the trailers searching for furnishings. Free stuff like that lasted no more than a few minutes.

My room was about a five-minute walk from the office, two if I ran quickly. There was some comfort in knowing that my place faced the prison and didn't get much foot traffic. But unprotected items sent out a proverbial dog whistle. The time between when my friend had left the furniture on the doorstep and my arrival would be only ten minutes. But that's an eternity in the furniture wars of Camp Bucca.

I darted through gaps and around corners, and finally sprinted down the elevated planked walkway that ran between the trailers. I leapt from the end of it and took a hard left, nearly knocking over an unsuspecting Iraqi worker. "*Asif*" – sorry – I said in Arabic, as I blew past him. I ran for the stairs and, seeing my hooch, sprinted for the door. To my disappointment, everything was gone.

Annoyed, I began the slow march back to the office. Sweating profusely with nothing to show for it, I kicked myself for not running faster. That was probably my only shot, I thought to myself. But then up ahead I spied a small shelving unit next to a door of what looked to be an empty hooch.

I could now see that it was a small bookcase, perfect for what I needed. A four-shelved system about three feet tall, made of cheap plywood, an ideal fit for the small space near the foot of my bed. The door was cracked just enough to barely see inside. It looked empty, as if someone was indeed moving out.

"Hello," I called into the opened hooch. "Is this giveaway stuff?"

No response.

"Excuse me, can I take this shelf?"

Again, nothing.

Although the room looked empty, it did seem strange that the bookcase could have lasted this long unattended outside an empty room. No matter. Questions could be asked later.

I grabbed the small shelving unit, heaved it over my head, rested it on my right shoulder, and began a brisk walk to my hooch. Sweat poured down my face now, as if someone were holding a garden hose directly over my head. But I was too pleased to care. I bounded up the metal stairs and into my room and set the plywood bookshelf in the small space between my bed and the wall. Perfect perch for my computer. Adding items to the shelf instantly gave me more floor space. Confident that my area was now officially functional, I stepped outside and locked the door.

Returning to the office, I noticed an airman standing in the middle of the long wooden boardwalk, outside the very room I had just visited, wearing an angry scowl. He looked around frantically, pivoting his head side to side, up and down. Then, suddenly, he raised the small broom in his left hand, swore loudly, and threw the broom down in disgust. It snapped against the boardwalk and tumbled a few feet away.

As I drew closer, the airman and I made eye contact. He eyed me suspiciously, and a look of recognition, tinged with anger, came over his face. It dawned on me then that he might have been in the process of cleaning his room. A recent sandstorm had blanketed the base, and microscopic sand particles had seeped into all the trailers, leaving a thin layer of dirt over everything.

As I neared, I could see him now second-guessing himself. Certain confrontation turned to indecision, and I decided to meet his uncertainty with a smile. I gave him a wave and some casual greeting. He opened his mouth to say something, but just as quickly dismissed his thought. I could now see clear inside the room, and it was not empty.

As I passed, I glanced over my shoulder once more and saw him still glaring, still contemplating a reaction. He just couldn't bring himself to do it. I briefly considered turning around. But doubling back now, to admit something so childish, would look even worse. I minimized the whole situation and pushed it out of my mind. And once I turned the corner, I sprinted for the office.

★

"You just took it?" Erin said.

"It was left out," I protested. "That's life." I crossed my arms defiantly.

"You're dead inside, Dave."

"He's right, though." The security forces captain had overheard our conversation and poked his head through the open door. "Can't leave that stuff out. That's just plain stupid."

The lieutenant echoed his sentiments. "He's a security risk. I mean, if he can't manage something as unimportant as a bookshelf in this environment, he shouldn't be in possession of a gun."

"I left for a deployment one year, came back stateside, and my whole office had been raided," said the captain. "They even took the pencils." He paused for effect. "I mean, who the heck uses pencils anymore?!?"

We all chuckled, except Erin. "So he's responsible because you guys have absolutely no integrity?" she asked.

"He knew the risk," I retorted.

But I wondered: Had I just stolen a bookcase? But someone had stolen mine. Small things like this did not register on my sliding scale of compassion. Bucca's furniture wars had turned me into a calloused soul.

CHAPTER 13

A Knock at the Door

Erin and I came through the gates just as a light breeze was working its way through the prison. The sun was calm and inviting. The weather was unusually pleasant.

We had agreed early on that we needed to get with Joe about the intelligence coming out of the prison. Now, with processes and procedures largely in place, we prepared to take advantage of the Al Qaeda intelligence only walking distance from us. Reconciling our intelligence with the prison would only strengthen its quality, perhaps providing new insight for policymakers seemingly paralyzed by the challenge of finding the roots of instability and violence plaguing Iraq.

We were also responsible, in my estimation, to be the tissue connecting the implications and consequences of the Sunni insurgency with the Shia-Iranian operations. We needed to be ready to speak with authority on the goal of handing over the Sunni-filled prison to what would likely be Shia management.

We had scheduled this rendezvous as our first attempt at bridging the intelligence divide. And while this meeting was merely to familiarize ourselves with the environment, we hoped that alone might open our eyes to new angles and inventive ways of thinking about sources.

★

The gravel path that led to the headquarters building lay ahead. My adrenaline pumped a little more than usual, but I played it calm as we walked down the path. Inmates on both sides ran for the fence, some shouting and smiling, others looking menacing. It dawned on me then that this was perhaps the only environment where enemies stood face-to-face outside of combat. A place where the rules of conflict didn't apply. Where law enforcement practices replaced war, although war still surrounded us. It was an awkward place to be.

Joe met us at the end of the path and ushered us through a side door. We entered a long hallway that made a hard left about fifty yards ahead and out of sight. Joe stopped us in front of a massive museum-quality display case that ran the length of the wall, full of shanks, knives, and other confiscated weapons ingeniously made by prisoners. "You guys got to see this stuff before we get the briefing," he said. One shank had a four- or five-inch blade made from the stem of a toothbrush, sharpened into a deadly point. A cloth ripped from a prison-issued shirt was wrapped tightly around the base to serve as a grip. A few strikes to the right area of the body would kill anyone.

"These things are a real threat," Joe said. "And the guards will get piss and dung thrown at them too. You get shanked with one of these prison-made weapons, you might get more than a stab wound." To illustrate his point, Joe launched into a story about one particular prisoner with a colostomy bag who would allow other inmates to use the hole in his stomach to simulate sex in exchange for money and other hard-to-come-by items. "We call him a colosto-tute," he deadpanned. Erin and I both nearly threw up.

We followed Joe as he turned and walked down the hall. The clean white walls looked out of place amid all the dirt and grime. Joe randomly shifted to Al-Qaeda. "Some bad guys are vicious, others just talk tough." He paused. "And you know, there are guys that are actually kind of cool. They'll chat, make fun of other Al Qaeda, all that stuff." His New York accent seemed more pronounced today. He explained that one high-level detainee spoke so disparagingly about Zarqawi that Joe initially believed it to be a ruse. But almost daily interaction with the prisoner revealed those feelings to be genuine. "He says Zarqawi is brutal and an idiot – I mean, dumber than a bag of hammers." I filed that little detail away and followed him through an open door at the end of the hall.

We stepped into a medium-sized room with white walls and a large, heavy wooden table that seemed to take up the entire space. This makeshift conference room had office chairs haphazardly strewn about the edges of a once proud mahogany table. Joe had set up a laptop at one end with a presentation for Erin and me to view.

As I squeezed into a chair that barely fit between the table and the wall, I pegged Joe with a question. "What was that random wooden outhouse at the fence line, right near the guard shack?" I had seen such an outhouse in each compound and it struck me as odd in its placement, nowhere near the inmates and unoccupied by guards. Joe looked at me, initially puzzled, then his eyes brightened with understanding.

"Oh, that's an isolation booth," he explained. "And it works like a charm." He explained that the inmates hated to be alone. Even the most heinous Arab men couldn't go long periods without physical touch. And I instantly knew what he meant. I had observed how cultural standards among Arab males, while arrogant and chauvinistic by modern American standards, were also much more reliant on physical contact with other men. I recalled seeing two Iraqi guys spooning in the shade just beyond the chow hall, as they watched a video on a cell phone. I frequently witnessed men holding hands in both Iraq and Kuwait. The cultural practice of physical touch among men had little to do with sex – although rumors of casual male intercourse were not uncommon.

"When they decide to play hardball, they get thrown in the box with only a Koran," Joe went on. "They start squealing pretty quickly after that."

The slide show on Joe's laptop was a briefing designed to familiarize a visiting general or media on generic happenings at the TIF, the Theater Internment Facility or prison. We flipped through explanations of the colors of prisoner clothing, the layout of the segregated camps, and perimeter security. That information proved helpful to our situational awareness, but it was of little use to our covert operations.

Joe switched gears, to an informal presentation on detainees that was more to our needs. He gave a quick rundown of the Baathists and Sunni insurgents currently housed at Bucca, confirming the existence of a rank structure within the Al-Qaeda areas of the prison, and explained how each prisoner was forced to obey a strict code of Islamic conduct within those areas, as a sort of rite of passage for terrorists. Zealot indoctrination occurred daily. And the prison's communal lifestyle allowed for near-constant interaction among high-level prisoners. As Joe explained, the environment created concerns that major Al Qaeda operators were not only recruiting and radicalizing other detainees, but also coordinating future operations. At minimum, Bucca was manipulating the casual observer into accepting the absolute adherence of Al Qaeda's apocalyptic version of Islam. Routine criminals were becoming hardened.

"As I understood it," I said to Joe, "we had decided some time ago to risk the bad guys radicalizing each other rather than have constant violence in the prison?" Joe nodded, conceding that there was no ideal option. "Violence in the prison is worse if we put them in camps with less violent detainees," he said. He was referring to forced recruitment or physical abuse of those who refused the overtures of the more radical. "And we definitely can't put Sunni insurgents with Shia; they kill each other." Ironically, though, the decision to group hardened Sunnis together would help bring into being one of the worst organizations ever witnessed: ISIS.

Joe wrapped up the presentation with a different set of slides addressing security information, several images from multiple angles showing an underground tunnel extending several hundred feet into darkness. "Guards discovered it after some detainees didn't report in one evening," Joe said. He glanced over at us. To me the shaft appeared tall and wide enough for men to walk through. Joe torpedoed that notion, claiming it was not as wide as it appeared. But a tunnel under Bucca was impressive nonetheless.

"If you look right here" – Joe pointed to a close-up image of the wall – "see how smooth and firm that looks." Erin and I both leaned in. I hadn't thought about why the tunnel walls were flawless, like the inside of a cannon. Working with tools and fingers would have made gouges and all kinds of divots. I felt dumb for not noticing. "They were slapping their tea against the sand to stiffen the wall," Joe explained. "That stuff turns the sand into cement."

In Kuwait during rainstorms, the wet sand would collect on the soles of my boots and harden within minutes. I literally had to use a bowie knife to scrape it off. Joe nodded in agreement. "That tea works wonders for getting the tunnel constructed quickly." It dawned on me then that the cultural dependence on tea meant these prisoners presumably had access to tons of it. Joe pointed to the dark abyss at the end of the tunnel. "It extended almost to the very last wall of concertina wire. A few more feet and they would have probably been able to escape."

His statement brought me back to the contractor's grim warning that my hooch acted as the last line of defense between the base and the prison. I had never once considered that the exit hole of a tunnel might show up near the door of my own living quarters. Erin looked over at me and gently placed her hand on my shoulder. "Good luck, Dave," she said. "I hope you sleep with your gun loaded."

"I'm not worried," I countered. "I'll take at least one of those knuckleheads with me. The report will definitely say, 'There was evidence of a struggle.'"

★

The room was pitch black, as was the world outside. I lay motionless in my bed, fading in and out of consciousness, as the muffled sounds of the distant prison echoed faintly. A glare from the floodlights in the compound seeped through a broken window shade and cast a glow across my face that forced me awake. My watch told me it was well past midnight. I reached over, flicked the shade back into place, and adjusted into a new position.

I had slept no longer than ten minutes, when I was awakened by a light thumping on my door. I propped myself up in bed and listened intently. The sound disappeared as quickly as it had come. I returned my head to the pillow and retreated back into sleep. A few minutes later I awoke to a louder thumping, like a child's fist rapping on the door. I rose out of bed and reached for my rifle. But it stopped again. I paused, motionless. I sat on the edge of my bed now, refusing to move. Nothing stirred outside, and for a minute or two it remained silent. I began to think I was going crazy. But just as I was about to return my rifle to its resting position near the edge of my bed, I heard several more soft knocks on the door.

No one would be knocking like that this time of night, and certainly not in an emergency. My mind returned to the tunnel, and to the potential for escaping inmates. Perhaps they believed they could lure me out this way. Breaking in would get them shot, of course. I tried to rationalize through it, practically shaming myself for even entertaining the thought of a tunnel exit in front of my hooch.

The knocking started again. Aiming my rifle at the door, I slowly turned the handle with my free hand and swung the door wide. Anyone standing there would have been knocked clear off the perch. But no one was there. I scanned from one side to the other. Empty.

I closed the door, now thoroughly confused. This is why I need rest, I thought. But almost immediately, the knocking began again. This time I lunged for the door, pulled it open, and jumped outside. Yet again I opened the door to a still, quiet night. I stood there dumbfounded. This time I scanned my area a bit longer. But even after my eyes adjusted to the distant objects in the darkness, nothing moved.

I was about to take a walk around my hooch, just to be sure, when I heard gunshots off in the distance. A training exercise had begun at the base shooting range several hundred yards away.

I discovered that evening that the sounds of gunfire echoed off my door so precisely that it sounded like a soft knock. Circumstances could not have been more perfect for such a revelation. Sleep would now be elusive. I packed my things and headed to Bucca Square.

CHAPTER 14

Throat Punch

I slept through breakfast this particular morning and by lunchtime found myself famished. Erin and I usually indulged in a large breakfast, which meant lunch was unnecessary until afternoon. But this morning, I met her near the entrance to the massive cafeteria at precisely 11:00 a.m., when the doors opened, and just before the long lines formed.

The vaunted Bucca chow hall, a wide expanse like an aircraft hangar, could fit several hundred people comfortably. Metal picnic tables covered an acre of floor space dotted with beverage, bread, and dessert kiosks. A U-shaped cafeteria serving line took up the front third of the building. The rest was roomy and open. The bright lights that hung overhead reflected harshly off the ceramic flooring, almost like a Hollywood film set. High ceilings and walls were lined with aluminum and metal siding.

In some ways, it resembled a local dive bar hosted in a barn. Military paraphernalia were scattered haphazardly along the walls, and banners of all sizes and colors sporting America's college and professional team logos dangled from the rafters. Some new, some ratty, some long, some short. It echoed like a drum, as metal clanked on tables and voices mixed together in a cacophony. But the cool air from the overworked AC units was soothing. It calmed irritated and sun-weary skin. Workers pushed food trays down the aisle. Aromas of cooked beef, baked bread, and steamed vegetables filled the air.

One of the chattier Iraqi linguists on base spotted us in the chow hall just after we sat down. I was looking forward to devouring my turkey sandwich, and he would likely ruin that.

As best I could tell, Yani was a low-level worker who assisted with minimal translation for some obscure office in division on base. He had no clearance and seemed at times just to be wandering around Camp Bucca. He engaged in lengthy conversations with anyone who would listen, leading many to assume that he had nothing to do. He was a short, wiry fellow from some town just south of Baghdad. He sported a full head of jet-black hair, spoke great English, and had a flirtatious personality. Sunken cheeks and a skinny frame made his bright blue eyes seem bigger, more interested. He had a reputation as an excitable character, and a habit of suggestive behavior toward women. Having to endure his company fell somewhere between entertaining and annoying.

Yani also had an air of duplicity about him. Whether it was his unabashed promiscuity or his complete lack of moral conviction about anything, I couldn't tell. But his pattern of haggling over everything, from the mundane trading of food to life-altering philosophy, made him come off as the sleaziest of salesmen. He had worked for some time in a bazaar that serviced American forces in Baghdad, which fit his persistent personality. In fact, that was where U.S. forces had found him speaking English. But something else, something more sinister, lingered near him like a shadow.

"Don't make eye contact," I told Erin without looking up from my food. "Yani just walked in." Always the devil's advocate, Erin turned to me. "Remember, Dave, everyone is a potential source." The annoying thing was that she was right. But before I could counter her comment, Yani saw us and made a beeline toward us with his tray of food. "Looks like he plans to stay a while," I said under my breath.

Yani rushed the table and halted just to the right of Erin, who sat directly in front of me. "My friends, my friends, may I join you?" I changed my countenance quickly, grinning widely and pointing to the open spot next to Erin, then leveled a patronizing smile in

Erin's direction, knowing she preferred space at the table. "Remember, Erin, everyone is a…" I began softly. "Shut up, Dave," she said in a hushed and defiant tone.

Yani sensed the fleeting tension. "Did I miss something?" Erin and I just stared at one another a moment, then I quickly changed expression. "No, my friend. I just like to give her grief whenever I can." Yani laughed and shook his head. "You are a character, Mr. David." He elbowed Erin. "How do you put up with this guy?" She just shrugged.

While organizing the food on his tray, Yani launched into his routine of complaining about the meal and retracing the previous day's activities, all the while working in a few sexually suggestive jokes. As always, he apologized to Erin mid-sentence as if he couldn't help himself. I wasn't fooled by Yani's banter. Unlike most talkers, Yani was also a good listener.

I noticed that he had three sodas on his tray. "You like caffeine, Yani?" He shook his head with conviction. "Water makes you fat," he insisted, running his hand over his midsection. "I need this to be trim, so I don't go girl hungry," he said between chuckles. "Sorry again, Ms. Erin." She didn't even acknowledge the statement. And he didn't even notice.

In fact, Erin had stopped paying attention the minute he sat down. I didn't realize it, but she had been plotting her exit from that moment. Yani continued his mindless chatter for several minutes. I nodded here and there between bites of my sandwich, knowing he was gauging my reaction. Then suddenly, Erin stood up without a word, stepped back over the bench, and wiped the crumbs from her pants. Not even making eye contact, she curtly said, "I think I'm going to head back to the office." She then reached down to retrieve her tray. Her eyes lifted, and now I could see her pleasure at leaving me to suffer alone. I nodded, acknowledging her well-planned maneuver. I hadn't even come close to finishing my meal, so I wasn't going anywhere.

Yani briefly rose as Erin left, mimicking an antique courtesy. This guy is unbelievable, I thought to myself. The whole charade made me laugh. A pervert and a gentleman.

"Was it something I said?" Yani asked, as Erin walked away. There was a genuineness in his question that I had never encountered in him before. "She knows I am just a jokester, right? I can be, how you call it in English … hmmm …" He looked down as he searched for the right English phrase.

"The word you are looking for is *forward*," I deadpanned.

"Oh, I thought it was *un-appropriate*?"

"Oh, I see," I replied. "*In*-appropriate is the correct way to say it, and you are that as well."

He let out a hearty laugh. "I can always trust you to be very honest with me, Mr. David."

His transition to the next topic was flawless. "Speaking of that, Mr. David, what exactly do you and Erin do here at Bucca?" His demeanor changed in an instant, his words dripping with duplicity. I had to be careful now. Any hesitation, any look of surprise, might confirm his suspicions. He was studying my face, so I purposely brought my sandwich to my mouth and took an unusually large bite to conceal my facial expressions.

"What we do mean, like my job?" I said. "The one we've talked about at length?"

He was momentarily on the defensive. But quickly he recovered and leveled more probing questions. "Yes, security for the base. But that's not all." He waved his hand dismissively. "I mean, Mr. David, let's be honest, you do more than security, I believe." He seemed almost angry now.

I laughed heartily at the suggestion, and his face slowly contorted into a look of insult. His eyes conveyed a sense of entitlement, as if I had to disclose the truth because he had had the audacity to call me on it.

"Is this the part where the Iraqi believes every American is a spy with the CIA?" I shook my head in amusement. "Come on, Yani. If I was a spy, I wouldn't bother talking to you." His eyes went wide, then squinted ever so slightly, the gaze of a man peering deeper into a statement to discover the truth.

He then became agitated – whether by my cavalier attitude or by his own inability to read me, I couldn't tell. "Okay, Mr. David, but I'm not dumb." He wagged his finger. "I believe you do more. I know you do more here." His now defiant tone gave me pause. There was something else there.

For now, I maintained my casual demeanor. "Yani, you can believe anything you want, I don't really care. I make sure our guys don't get killed here, and then I go home. Don't wear yourself out with conspiracy talk. It's exhausting."

Yani threw up his hands in disgust, a visible outburst that surprised even himself. Quickly he regained his composure and grinned, as if to acknowledge that we both had seen that lapse in calm. "Yes, Mr. David," he said through a precarious smile. "But I still think … I still think there is something more."

★

I was fuming inside as I walked through the office door. Yani had had the nerve to test my ability to keep my cover, and to be angry when I didn't crack. We faced bigger issues than fending off inquisitions from a low-level nobody. That arrogant thought reminded me though that, whether a worthy opponent or someone like Yani, I could not underestimate any enemy. The whole charade was infuriating, but also a valuable occasion for self-reflection.

"How was lunch?" Erin asked sarcastically.

"Well, I want to throat-punch Yani," I told her. "Is that good enough?"

That earned a healthy chuckle. Erin had a winsome laugh that could ease a stressed soul. As I grew more relaxed, I was better able to recount the conversation that had transpired in her absence. "Well, we know he's stretchy," she said. "So does this surprise you?"

I considered her question for a moment. "But there was a determination I've never seen in him," I replied. "He acts like a playboy, not really interested in the war. Something about it was very unsettling." Erin looked up from her paperwork and stared forward, as if processing my comments. "You think we could flip him or have one of our sources get close to him?" She was reading my mind. "I'm going to have one of our guys get info on him – smoke some hookah with him. I'm afraid he's working for somebody."

★

It was now late afternoon, and the weather had cooled to 90 degrees or so. I was plotting who to run against Yani, contemplating the best option from our limited stable of reliable sources, when the ring from Erin's desk echoed across our tiny space. Erin picked up the receiver and wedged it between her face and shoulder as she continued to type away on the computer. I heard some casual chatter on the other end of the line, then she looked my way. Pulling the voice end of the phone from her face, she said, "Dave, they need you at the front gate. Some guy says he has info." She brought the phone back to her mouth, mumbled a thank you, and hung up.

Walk-ins, as they are known, show up unannounced, ready to provide information. They can be incredibly helpful but are the most dangerous of sources. They could be genuine and self-motivated, a prized asset because they come equipped to work, rather than having to be convinced. Or they could be a "dangle" – a double agent posing as a legitimate source deployed to penetrate an adversary's operation.

"Take Josh with you," Erin said as I gathered my notepad and gear. The newest member of our team had arrived in country about several weeks prior, and his first deployment was proving more laborious than he had anticipated. He showed some hesitancy in adapting to my unusual calendar, preferring to maintain a stateside office schedule. I argued that deployments required round-the-clock creativity and flexibility. This unusual call to the gate during normal business hours would give Josh the opportunity to gain some experience.

"All right, let's go then." I grabbed my gun, and we made our way to the door.

The walk took about five minutes, enough time for me to prep Josh on the process, and we agreed that I would take the lead in the questioning. We approached the security center, which served as the headquarters for visitors processing into the base. Its commander happened to exit the building at the same moment, black shades shielding his gaze from the stupendous glare. As we approached, he stopped abruptly. I could not see his eyes to know if we were the reason. "You here to talk to the walk-up?" He apparently recognized us, presumably due to our beards and lack of uniforms.

I removed my sunglasses. "Yes sir. What do you think about him?"

The commander shook his head. "Haven't spoken with him. Some other guys are in there right now." He swung wide the door and motioned for us to enter. The space was incredibly clean, for a warzone, outfitted with the famous plywood desks and waist-high partitions. A few folding chairs lined the wall of the tiny waiting area to the left. A security forces officer manning a desk several feet directly in front of us looked up momentarily from his computer, and just as quickly returned to his work when he saw the commander with us.

We were directed down a short walkway. "Third door on your right, just knock on it, he's expecting you." I nodded my thanks and began walking that direction. Peering through a small window in the door as we passed, I saw a small room like an interrogation booth or holding cell, with just space enough for two chairs and a tiny side table. The third door opened just before we reached it and a large Air Force sergeant stepped out, nearly colliding with us. "You here to talk with the visitor, I assume?" he said, in a thick Minnesota-sounding accent.

"Before we go in, give me a little background," I asked him.

The sergeant explained that the local Iraqi worked on the base regularly, spoke excellent English, and had no weapons when he arrived at the gate. He claimed to have seen some suspicious characters near the perimeter during weekend returns home to Basra. Something about the men and their positioning along the perimeter of the base seemed odd to him, but he wouldn't elaborate. "I think he wants money before he'll talk," the sergeant concluded.

"Did he act shady? Any behavior you find strange?"

"No, he seems quite genuine actually, and even a little nervous when talking to me."

"All right, let's have a chat." I looked over at Josh to ensure he was ready. He looked content, so I nodded to the sergeant, who slowly opened the door to the small room.

There sat a balding, heavyset man who immediately straightened in his chair when we saw us. He had a wide mouth, and his eyes seemed friendly. He looked to be in his late thirties or early forties and sported a brown polo shirt with what had been a thick white stripe across the front, though years of wear had turned it yellow. His tan corduroy pants appeared to have gone unwashed for months, but he wore the customary dark brown sandals. Instantly I noticed a lack of body odor. My senses had begun to anticipate the smell before any encounter, and its absence was curious to me.

The gentleman now turned in his seat, looked at the three of us, and vehemently shook his head. The sergeant spoke first. "These are the guys you want to talk to. They'll appreciate your information." But the Iraqi shook his head again, in a way that suggested he understood the sergeant but had other reasons for his refusal. He then pointed directly at me. "I will only talk to him," he said. I tried not to show my surprise. We calmly agreed to the request, but I explained that my partner would have to wait just outside the door. He nodded in agreement and turned away. I took that moment to adjust my handgun to a better position in my waistband and left the door cracked.

Then I took the seat directly in front of the man. Only two or three feet separated us. I evaluated my seated position and adjusted my posture in a way that would allow me to strike quickly, in case he decided to lunge. He didn't seem to notice my guarded posture or care that I remained defensive. And before I could speak, he explained why he had asked for me. "I'm sorry. I don't mean to offend your friend, but I believe I can trust you. I know you care about Iraq and the people here." He paused. "I see you out with us, at hookah, and I know you are a good man …" His voice trailed off, as if he were embarrassed by his candor.

I was taken aback. I had never spoken with this man, nor did I remember ever seeing him at Bucca Square. Yet he watched me regularly and took notice of my interaction with the locals. Naturally, I was suspicious. But as he unloaded his story, I began to believe him. Months of late-night rapport-building had apparently paid off.

Ramsey proved to be a gentle soul with a sharp mind and a keen interest in the war effort, although he had limited access to the information we needed. He would become an irregular contact of mine.

Between Ramsey and Yani, we knew without a doubt that we no longer had the anonymity we had once enjoyed. The following two months would move us from development of sources to execution of operations.

CHAPTER 15

Arousal

By late May, the sectarian violence had reached a fever pitch. It seemed that the warring factions, especially the Sunni extremists, had an insatiable thirst for blood. AQI and its affiliates had unleashed an inconceivable level of violence. We found a lot of useful information from workers who traveled weekly to and from Baghdad, a six-hour drive north of Camp Bucca and outside the Shia stronghold of southern Iraq. These sources returned with disturbing news of Sunni hit squads seeking out Shia families for execution, both random and calculated. The initial tales of masked men storming into homes in the dead of night to execute or abscond with family members had grown increasingly sinister as torture, rape, and other brutal tactics replaced the standard snatch-and-grabs. The dress of the intruders, their demands, and their weapons all pointed back to Zarqawi's Al Qaeda in Iraq.

From my vantage point, it seemed the bloodshed here had ratcheted up the political infighting in Washington. I watched from afar as partisans passed off the worst examples of their opponents as the majority opinion. News media, those who rarely do well with nuance, tried to explain the violence from one point of failure. But doing so ignored the complexity of it all. The civil war label grew in popularity, but the idea of competing parties positioning themselves for governance – the traditional definition of a civil war – didn't quite fit the situation either. As we say in the military, people were trying to square the circle.

It remained the job of the intelligence community to slog through the minutiae of it all to create a truer site picture. We were expected to deliver facts, rather than politically acceptable spin. We had no responsibility to provide data that fit an ongoing narrative of either the administration or its foes. We were responsible to transmit intelligence, even if it described a disappointing reality. And much of what the leadership wanted to know had a connection with Camp Bucca. It felt like the center of the universe.

★

We had spent most of the day outside the wire, shadowing patrols as they conducted perimeter checks along the fence line. We encountered a mob of local teenagers hawking old Iraqi currencies as we searched for indications of explosives, lookout posts, and sniper pits. The money was no longer in circulation, and these were likely counterfeit. Yet I thought they looked like an interesting collector's item, so I pulled out American cash to purchase a handful of the bills bearing Saddam's image. "That's a mistake," said a sergeant as I reached into my pocket. Within seconds, a herd of young men swarmed our position. I nearly had to use my gun to back them off. The chaos that ensued endeared me to no one on our squad that day.

Darkness had descended on base by the time we returned to the office. We hastily discarded the heavy gear and racked the weapons. I fell backwards into the chair, exhausted, when a call came through on Erin's desk phone. I was the closest to it, so I used my heels to scoot the wheeled chair over to her desk.

"Is this OSI?" asked the voice on the other end of the line.

"Yup, what's up?" I pushed the receiver tight to my ear, as people continued dropping gear to the floor, noisily echoing off the walls of our hollowed-out plywood office.

"Your command is calling, says it's an emergency. How quickly can you get over here to take the call?" Puzzled as to why OSI hadn't called our phone directly, I instinctively replied that we were on our way. "Guys, command is calling, says we got an emergency. They're on the line at the comms building."

Erin was first through the door of the communications trailer, which was only about fifty yards from our office. I had never been in the base nerve center, and was pleased to see how professionally it was put together. Wires crisscrossed the floor in tight, organized bundles, all tied to phones and computers themselves neatly segmented on several desks. Maps were spread across a handful of long tables, and a massive white map of southern Iraq took up almost the entire back wall. It showed an intricate layout of terrain differentials, roadways, and pockets of urban areas amid desert. Routes had names; red pins marked specific locations.

The room was a beehive of activity. Dings from incoming emails, telephone ringers, and the sound of people furiously typing away at their keyboards created an atmosphere of constant movement, progress toward an unknown goal. Thankfully it was already dark, and the crews were in the process of shift change, so the room wasn't as packed as it might have been during the daytime.

"We got a call that OSI command is on the line?" Erin asked quickly, looking around the room. A soldier standing near the massive white wall map waved us over to his landline phone. Erin grabbed the receiver and listened. She motioned for a pad of paper, which I dutifully handed her. She grabbed a pen from her pocket and began furiously writing notes. Her face registered an uncertain, almost skeptical expression. Whatever the conversation entailed, Erin was not sure she agreed. "All right, I have the team here. Hold for a second and let me get their input. Be right back." She put the phone down and turned to face me.

"Apparently a police station has been overrun near the base," she said. "Whoever it is has apparently killed or kidnapped some people close to a source of some other agency. They want to know if our sources know anything, and if we can get a better picture on the situation."

"You mean like get eyes on the police station?"

"Sounds like it."

We exchanged confused looks. It sounded like someone in OSI or another agency had a source who had contacted them for help. Erin returned the receiver to her ear. "I'm back," she announced, then began jotting down more notes.

I turned to the Army sergeant across the table. "How quickly can QRF be scrambled?" He explained that the quick reaction force would take only a few minutes, but we would need prior authorization from the commander, since the alleged attack didn't involve U.S. military.

"I figured as much," I said. "Are there any assets scheduled to leave or outside the base right now?" He turned his attention to a map and shook his head, without looking up. "And no patrols scheduled to leave soon?" Nope. "And I assume no reports of a violent takeover nearby? Gunshots, nothing like that?" A rhetorical question that received little more than a blank stare.

He glanced over at his subordinate manning the hotline, who had turned his chair around and was listening intently.

"Anything?" the sergeant asked him.

"Nothing, Sergeant. And we would hear about something if it was close."

I then quickly spun around to our linguist. "Get our guys on the phone, find out if they know of a police station takeover nearby." He hurriedly grabbed his cell phone from his breast pocket and began thumbing through his call list. Within moments, he had one of our guys on the phone. "Take it outside," I whispered. I trusted our military brethren and had no reason to believe they could understand Arabic. But protecting the source had to be a universal practice. I opened the door for our linguist and ushered him outside into the night.

Erin yelled for me to join her at the phone bank. I jogged to the back of the room, catching the look of exasperation and uncertainty on her face.

"They keep suggesting we get eyes on. They're pushing us to find out more." As we had suspected, some source had phoned it in, and our command wanted to capitalize on their new agents at Bucca. I turned to the sergeant. "Are there even any police stations close by?" He pointed to the map on the wall, fingering a few locations to the southwest and the north. "But they are a good five-to-seven-minute drive after you leave the base."

He looked skeptical that anything of this nature could have happened at a nearby police station. I was growing increasingly skeptical as well. "They don't have any more information? They know we don't have the capability to respond to this, right?" My question was for my own benefit, said in frustration since Erin knew the answer. "Sounds like HQ is pushing Kuwait to respond, so that's why they're calling us," she said.

"Well, let's saddle up and go check," I said abruptly. Another agent responded enthusiastically, and I knew immediately that my comment had been ill advised. It had also caught Erin off guard. And before she could respond, our linguist burst back through the door. "Nothing. Nobody has heard anything," he said breathlessly. "Our guy in Safwan just laughed at me. He said all the cops go home; nobody likes to work at night." That wasn't entirely true, but he had a point. The likelihood of this incident even having occurred was diminishing quickly.

We had now been in the emergency center for about ten minutes. The window of opportunity was closing quickly. "Let's talk outside," Erin said in a hushed tone. We stepped through the door and onto the short metal staircase that led into the trailer. The air felt especially cool. Nothing stirred. Although bright and decision-oriented in her own right, Erin often conferred with me, as her fellow officer, about major decisions. But tonight, she scrutinized me through squinting eyes.

"Dave, are you serious?" she said, nearly pleading with me. "Do you think it's wise to jump in the truck and run out there without any good knowledge of where or who or what the heck is going on?" She gave me no time to answer. She swiveled her head and swept something off the step with her foot. "For all we know, we're walking into a trap. We don't know what kind of firepower this group has, where they're at, if they even exist." She made eye contact again. "I mean, do you really think we should roll out?" Since we generally were on the same page about strategy even without consulting each other, I could see that my comment had stunned her.

"No," I said flatly. My quick and blunt response startled Erin. A look of surprise, then relief, spread across her face. "That was a stupid comment. My testosterone got the best of me." Her shoulders visibly relaxed. "Sorry, bad move on my part."

Before she opened the door, Erin looked back at me. "Dave, we're friends. And we will always be friends, right?" A gust of wind kicked up just as she spoke and carried with it a slight whiff of body odor. I suspected the laundry folks on base of merely drying, not washing, my sweat-stained clothing. Week after week, my shirts came back smelling awful.

I looked back at Erin. "Of course. Why would you even ask?" She looked down at her shoes for a moment, then up with a wincing smile. "Because, Dave, you stink. I mean, really bad." I hung my head, and she was back through the door before I could respond.

Erin called back to headquarters and strongly recommended against trying to get eyes on. We waited as our opinion made its way through to the Kuwait office. No one said anything. Erin stood with the receiver pressed between her ear and her shoulder, smoothing her shirt as a nervous tic. I occasionally glanced at the map, then over to our young airman handling incoming emergency calls. He sat silently, saying nothing, monitoring his computer screen.

Erin looked at her feet and took a deep breath. "Okay, yes sir." She hung up.

"I think they need a serious lesson in our capabilities here," I said bitterly. "Do they know we're leashed to the doghouse?" Erin just nodded. We had explained our situation on more than one occasion. This latest request proved that our people still did not fully comprehend our limitations.

★

The next morning, we fielded a call from our superiors in Kuwait. Though supportive and even-tempered, they were anxious to know why we had not responded to the supposed overrun police station. Erin and I explained, as calmly as we could, that the information had been scant and unreliable. Moreover, we did not have a team large enough for security, nor did we have QRF available that might have allowed for an immediate response.

Revisiting our circumstances was routine. But defending against perceptions of inaction bordered on demoralizing. The mere process of justifying our refusal to act was hard to do. We prided ourselves on accomplishing tasks, no matter how impossible. There was no escaping the sinking feeling that someone would interpret this as us doing just the opposite.

Our bosses meant no harm. As the conversation continued, it became clearer that they would use this instance as evidence to build a case for their superiors on how the limitations we faced hampered the mission. It was clear Air Force leadership took a keen interest in our work. It was also clear they still did not fully appreciate our circumstances. The cruel irony of the whole situation was being in an environment that prized action, even as we were saddled with rules that bred inaction.

I took comfort in knowing that our small team was gathering an impressive amount of information proportional to the capabilities and freedom of other units. And regardless of whether anyone noticed, I could sleep easy knowing we worked hard and did the right thing. "Well, that went better than I thought it would," Erin said. We sat in silence for a moment, fatigued at having had to defend ourselves. Little did we know that fatigue would soon be tested again.

★

Army investigators called in a near panic: "We need you guys over here now." Someone had apparently absconded with one of the waste trucks, a refurbished cement mixer that rumbled around base each day, cleaning out the contents of the port-a-potties. An Iraqi worker usually manned the vehicle as it stopped at each location. He would extend a long plastic tube resembling a fire hose from an opening on the massive mixer, then throw the business end of the hose into the base of the outhouse and suck out all its putrid matter. The smell was stomach-turning. By the end of the day, these brave souls would have traveled the entire base and collected all its refuse.

Everyone knew the implications. A waste truck loaded with explosives could take out half a city block. The base commander was frantic for answers. We joined others in security as we raced from office to office trying to locate the workers and contractors responsible for the machine. Subsequent interviews yielded painfully little information. Erin and I were genuinely dumbstruck that no one had seen it leave. I knew that it had to be an inside job. I had experienced something similar in Afghanistan. But, for now, it seemed the truck had vanished into thin air.

The Marine commander called the meeting to order in a controlled rage. To no one in particular, he thundered: "How can a massive dump truck just disappear!?!" The room was silent for a moment. Then, one by one, unit directors did their best to announce their findings in a calm, deliberate manner.

"The driver from yesterday said he just left the keys in the truck, like he does every night," the lead contractor said.

"The gate guards do not remember seeing a truck exit during the night," an Army captain added. "So it probably happened right after daybreak, likely when all the semis are making their deliveries."

Erin spoke next: "No one was seen near the truck this morning. And no one has been identified as missing from among the local workers."

The commander flung his head backward in disgust. "Unbelievable. Simply unbelievable." I began to speak in a shaky voice. "This stuff happened at Bagram Air Base in Afghanistan several times." The commander slowly brought his face level, revealing a look of horror. I tried my best to ignore his expression. "Toyota 4Runners were stolen off the base regularly," I told him. "I mean, theft is not uncommon throughout the warzone." Quickly, I wondered why I had said it like that. I had intended to imply that it could have been an inside job, but I had absolutely no information to back that up, and I stopped mid-thought.

A few sideways glances made me squirm a bit. I thought I saw a flicker of relief in his eyes, perhaps in knowing that the theft of government equipment would come as little surprise to his superiors. The meeting moved along, and the real question became finding the truck before it fell into the wrong hands.

The commander for the Army investigative unit surprised Erin and me by insisting that OSI conduct a formal investigation into the matter. The room fell silent for a moment. I could feel the fatigue setting in, as I anticipated having to defend our refusal of yet another request. We told Air Force command our limitations, and still requests came in that far exceeded anything we could do. We had already explained to Army investigators as well that we were equipped for and tasked exclusively with counterintelligence work. Even if we wanted to participate, we did not have the tools for labor-intensive witness interviews, crime scene analysis, and case preparation.

"We do not conduct criminal investigations in the warzone, unless strictly advised by OSI command or tasked to a unit dedicated to criminal work," Erin said, as if reading from a Department of Defense instruction manual. Our sole mission remained counterintelligence and counterterrorism, she advised. The Army commander was having none of it. "Even the possibility that the truck will be used as a bomb made it a counterterrorism concern," he said.

The logic baffled Erin and me. There was simply no evidence to suggest that it was related to terrorism or counterintelligence. There was an equal chance that thieves had taken it and sold it to a cement company that would later win contracts with the U.S. government. There was as likely a chance the truck had already shipped to a neighboring country. That was a valuable construction vehicle.

There were more people who would pay good money to use it, and fewer that would pay top dollar to destroy it. In my brief experience, stealing materiel from the Americans was far more profitable if end users didn't use that equipment against the Americans. Theft was often just that: garden variety theft.

I could tell the Army commander wanted to push this off onto someone else. And we balked at the absurdity. Our office couldn't spend time chasing down imaginary scenarios, especially when the base had its own investigative unit across the table from us. If their investigation found a counterterrorism nexus, Erin offered, we would take it over immediately. To move the solutions along, we added a recommendation that the base issue a "BOLO" – "Be On the Lookout" – for the dump truck to all our units across the country, with as much detail on the vehicle as we could ascertain. We volunteered to leverage all our sources to locate the truck.

Erin recommended that the two offices hammer out the details privately. The base commander nodded and dismissed us, clearly wanting answers and not caring who provided them. Erin had defused a difficult discussion. We reiterated our position to an angry Army commander in a closed-door conference a few minutes later. We were not lazy. We just had a different job to do, she explained, with her famous steely-eyed calm.

Looking back, I have little doubt that OSI's policy utterly confused the Army. They had a separation of powers divided between criminal and counterintelligence. We had no such partition. OSI did it all, and the mission changed depending on the location. That apparently gave them the impression that we could transition back and forth between specialties at will.

"Well, they're not happy with us," I said to Erin as we returned to our office. "I'm surprised they wanted us to sidestep our mission. They would never have done that." Erin just nodded in agreement. I was emotionally exhausted, disappointed that we had to turn down another opportunity. Saying no went against every instinct we had. And yet, again, it was the right answer.

She had taken the brunt of the hostility. She looked on edge. From fighting for office space to explaining our mission ad nauseam, the never-ending saga of internal disputes was wearing us down. And it was all made worse by the perceived lack of external support.

I knew that intelligence work required patience, stamina, and perseverance. I was learning that life required that too.

★

"I need some laughter," I said to Erin as I fired up my computer. I scanned emails with little real interest, flicking through, channel surfing my inbox. One had a threatening tone in the subject line that led me to pause and open it. It was from the first sergeant on base, warning everyone at Bucca about the dangers of using aerosol cans in the rooms. "These types of products are combustible," it read, "and have been known to cause serious injury when left in confined spaces in extreme heat." The brief instruction ended with a prohibition against their use inside the hooches. The problem was that *aerosol* had been spelled *arousal.* Base leadership had just outlawed the use of "arousal cans."

The error threw me into fits of laughter. "Erin, read the email from the senior," I said between gasps. Before she could even find the message, our security forces neighbors let loose with a chorus of booming laughter that nearly shook the trailer off its foundation. They were now banging on the plywood walls: "Guys, where the heck can I get an arousal can?" Erin read the warning and looked more shocked than amused. Another call from the other room. "I need one of those. I've been here for eleven months!!"

The laughter subsided, and minutes later Abdul, our linguist, burst through our office door. "I'm glad you're here," I said. "You heard about the truck? I need you to contact our guys about …" But before I could finish, he cut me off. "Listen, David, one of our contacts in the police just called and said to never leave the base at night. He said people are trying to get soldiers off base to kidnap them." We listened intently now. "You mean, like fake a police station takeover?" I said somewhat sarcastically. Abdul shrugged. "He didn't know how they get you off base, but he said he knows they are trying."

If only for a moment, we felt vindicated.

PART THREE

CHAPTER 16

A Lion Cornered

Our work picked up speed in those final days of May. Our sources uncovered Shia militia using highways near the base to ship Iranian heavy weapons north into Basra and beyond. We had times of delivery, vehicle descriptions, and precise routes. We hadn't identified the final destination, and someone in Baghdad probably hadn't identified the origins of the weapons. But that's how it worked. We didn't need to know their work, or vice versa. We only needed to publish intelligence and let strategic planners put the pieces together.

The same could be said of the prison. The detainee network of intelligence proved that Baathists and former military and intelligence officers in the Saddam regime, like Abu Ali al-Anbari and Abu Muslim al-Turkmani, played vital roles in prolonging the Sunni insurgency from inside Camp Bucca. They were limited in action, to be sure. But recruitment, indoctrination, and networking created an informal system for expanding the organization, once prisoners saw freedom. Lesser known prisoners, like Abu Bakr al-Baghdadi, who would later become the leader of ISIS, had rejoined Al Qaeda forces in Baghdad after their release. Author Abdel Bari Atwan would later refer to Bucca as a "jihadist university."

All this work had caught the attention of more and more war planners. Outside calls, much like the alleged police station takeover, were arriving with more frequency. We were being pulled deeper into the compartmentalized sphere of intelligence sharing, and that only proved that our information had value and that users wanted to share it.

It all came at an important time, since the Pentagon was demanding that military intelligence agencies partner in more sophisticated ways. But for us to be successful, Erin and I knew that, at its core, intelligence sharing had to be a willingness to leverage our network for information that might have little influence over our area of responsibility. We had to become more accustomed to receiving requests that were unrelated to any mission or operation in which we were currently involved, and to accepting those requests knowing that some senders were unable to provide reasons or context for the questions. To the novice, these stringent "need-to-know" restrictions on dissemination appear harmful. But we knew that, on the contrary, quality information demanded tight controls over how it was gathered, analyzed, and distributed.

The call for continued intelligence sharing had to do with an overarching interest in finding Zarqawi. The hunt for the elusive Al Qaeda leader proved so important that it turned into a daily presidential briefing. We contributed information as best we could, although we had no doubt he was holed up somewhere near Baghdad. The noose was tightening around Zarqawi – and, like a cornered lion, he fought back.

★

Erin banged on the door to my hooch before the sun had even risen over the prison. Startled, I dressed quickly and swung the door wide to find her standing there. "We have to move. Soldiers were kidnapped in Baghdad." Without a word, I spun around and grabbed my gear. Within minutes we were walking through the door of our office.

An early morning phone call from OSI command told us that American soldiers had been kidnapped at a checkpoint somewhere in Baghdad. Initial details were scant, but the instructions were clear: leverage our network of sources to help find them.

"This isn't good, Erin," I said. "These guys don't come back alive."

She nodded and began typing. My monitor hummed to life as I moved across the room to our secure computer. I sat impatiently, as both took a painful amount of time to establish a connection. "We share intelligence; we just can't share it fast," I mumbled to myself.

After several minutes, I was able to pull down sensitive reports. Unfortunately, these yielded little additional information. Thankfully, reports began trickling in just as my patience was wearing thin.

We pieced together from multiple situational reports that two to three troops manning a small permanent checkpoint in a dense, urban area of Baghdad had radioed for backup after coming under attack. Soldiers were routinely deployed to these makeshift duty stations scattered around Baghdad, to serve as checkpoints and extended security. They were often far from any base, and in place one to two weeks at a time, alone in a congested area of the city. Those postings were dangerous work.

The responding units found this particular checkpoint deserted, save for spent shell casings and some blood, all of which pointed to a struggle. And it wasn't long before kidnappers from Al Qaeda boasted that the soldiers were in their possession. Evidence gathered at the scene had already suggested that armed militants likely ambushed the American checkpoint during the late evening hours and absconded with the soldiers. That information proved accurate.

We went to work tasking all our sources to find something, anything that might lead to their whereabouts. We also coordinated with Joe to see what the inmates might know. Al Qaeda prisoners almost certainly had information, at least regarding a hiding location, and we were prepared to help Joe interrogate each and every one of them, if it came to that. We would spare nothing to get these guys back.

Joe decided to go after the friendly, high-level detainee who had openly derided Zarqawi as an idiot. I knew from Joe's information that Makeel was a smart and deadly man who would have intimate knowledge of Al Qaeda's network of safe houses and strategies for hostage taking.

"Joe had already planned to meet with him today," I said to Erin after hanging up the phone. We waited several hours to hear back. It felt like an eternity. Joe didn't bother calling either. Instead, he showed up in our office after concluding his session with Makeel. "He's got nothing," he told us. "And he says we won't find them." Erin and I looked at each other, unwilling to concede defeat. But we had a sinking feeling that he was right.

As Joe told it, the senior Al Qaeda operative explained in no uncertain terms that his group had definitely taken them, or at least the tactics sounded like those of Al Qaeda. Makeel even told Joe how the mission had probably been executed. "He pretty much confirmed every detail of the attack, without me telling him what our forces found at the scene."

And where they took them was anyone's guess.

According to Makeel, Al Qaeda had safe houses spread all over the place, even immediately near the site of the kidnapping. He listed a few places where the soldiers might be but, between the lookouts and security, it would be impossible to get close enough. "They will move them every few hours," Makeel explained. "And if they sense Americans getting close, they will just kill them." Makeel said the only real possibility was to catch them in the process of moving the soldiers. But even that was a long shot, he admitted, since Al Qaeda was expert at concealed movement in urban areas. It was later learned that Zarqawi himself regularly traveled through Baghdad in all types of disguises.

If the rescue operation sounded implausible, a negotiation for them was impossible. Makeel said that Al Qaeda would make some outrageous demand, such as the redeployment of all American forces out of Iraq within one week – something the U.S. military could never achieve even if it wanted to. "When you fail at whatever they ask, they will kill them and probably dump their bodies for you to find."

We continued to leverage our sources, despite Makeel's information. Twenty-four hours later and still no news. Every search turned up nothing. Each lead was a dry hole. Most of those readily available to us on base were local Shias who rarely traveled north and had little contact with Al Qaeda types. Those that did have Baghdad connections had nothing or failed to answer their phones.

Joe circled back with Makeel in hopes that he would have remembered something – perhaps a location, a building. The Al Qaeda operative dismissed any thought about a successful recovery. "He pretty much said to give up," Joe explained grimly over the phone. "If we haven't found them yet, then our guys are probably dead."

Three days after the abduction American forces received an anonymous tip, revealing the resting place of the bodies. A search team located their discarded remains near the site of the ambush, suggesting that they hadn't been moved far – an absolutely discouraging revelation. Al Qaeda operatives had littered the scene with roadside bombs, making it nearly impossible to retrieve the fallen soldiers quickly. Even from a distance, though, first responders could see that both soldiers had experienced significant torture. It took a bomb disposal crew nearly a full day to remove all the explosives and secure the area. They soon discovered that Al Qaeda operatives had also booby-trapped the mutilated bodies with more bombs. It was several more hours before medical teams could safely retrieve the bodies.

Quickly I realized that American search and rescue teams must have scoured the areas around where they had been hidden. It was hard to accept that Americans might have been within blocks, even yards, of those soldiers without knowing it. My heart sank to imagine that our guys might have heard the rumbling of Humvees, the sound of search helicopters overheard. The families would be notified soon. Parents, wives, children would be inconsolable. They would naturally want to know more. But they would quickly learn that they didn't want to know the details.

★

I knew death. I had seen its gruesomeness and had watched the excruciating pain it caused. But familiarity did little to buoy my disappointment or assuage my anger.

But war didn't leave you much time to ruminate. A rocket attack on Bucca later that day served as a not-so-subtle reminder that more violence awaited us all. The explosion rocked the base, but the missile landed harmlessly near an open field along the perimeter. A few days later, some especially important people within the intelligence community arrived at Bucca unannounced. They headed to the prison without much fanfare and departed the base the same way. We knew that meant something important.

CHAPTER 17

The Big Salad

The transmission originated with another American unit near Baghdad. This particularly reliable intelligence team had picked up some information indicating that Bucca would be the target of an attack in the near future. I was less than alarmed. We routinely received intelligence warnings that pointed to imminent strikes of one kind or another. But this time, as I dug into the details, a story emerged that made me think differently.

Someone planned to attack the famous Bucca chow hall. The apparent strategy suggested someone not only with intimate knowledge of the base, but who knew the layout of the cafeteria itself. We had no information yet on the attacker or attackers, or the possible time. The source was close enough to the bad guys to know of the attack, but not so close to know the perpetrator. All the details in the report pointed to an inside job.

Before I could present this report, along with my hunch, to the base commander, I had to analyze the information, interrogate the report for more gaps, in preparation for the questions that would undoubtedly arise during the briefing. It was pleasing to see how the up-tempo intelligence process in Iraq had encouraged this type of sharing and transparency. Typically, another unit would be dissuaded from forwarding unredacted reports, since details could reveal the source or fall into the wrong hands. But in this case the sending agency trusted the process enough, and felt it to be sufficiently urgent and reliable, to forward it in its entirety.

But with speed and efficiency came raw and incomplete information. The resolve to stop a pending attack was more important than ensuring that every detail proved accurate. So the report had many holes. That left leadership in a quandary. They would have to decide how to adjust security and reallocate limited resources, based on partial threat information. The commander would know that rerouting security elements would leave Bucca vulnerable to attacks elsewhere. Regardless, consequences awaited a commander who did nothing with such intelligence if an incident occurred.

Meanwhile we felt the weight of this decision, along with a responsibility to interpret the intelligence and deliver the most reliable assessment possible. We wanted neither to overstate nor to understate the potential threat. I quietly worried that the enemy might have planted the information as a decoy, to elicit exactly that response and assault the resulting weak spot.

"This could be anyone," Erin said. "There are so many foreign nationals that come on and off this base. I mean, even the prison guard training." She was referring to the training camp for local detention officers, mostly Shia Iraqis, many of whom sympathized with the likes of anti-U.S. cleric Muqtada al-Sadr and his Iranian supporters. It was a gold mine of intelligence, but also a constant worry.

The intelligence was rather clear that someone on base was serving as the main player in this possible attack. We felt confident in that conclusion. Intelligence requires agents to operate with the presumption that a mole exists, with or without undeniable evidence to prove it. It sounds paranoid to the untrained observer and, admittedly, counterintelligence can wear on one's sanity. But history teaches that assuming an inside threat doesn't exist, because you haven't found concrete evidence of one, is a perilous strategy in the world of spy craft.

Before Erin could offer more, I turned to Abdul, who was sitting patiently beside me. "Get on the phone with our guys here …" But before I could finish, our eager linguist tried to complete my thought: "… and find out if they have heard anything. Yes, yes, Mr. David."

I waved off his comment. "No. They should have called us by now if they had heard something. Otherwise they're worthless." Abdul conceded the point with a nod. Before he could put the phone to his ear, I grabbed his arm. "Ask them *who* they think might be capable of hurting the base. I guarantee they're already suspicious of someone. So who is it?" I figured we could tackle the problem from a different angle, one that would allow us not to reveal everything we knew and still gather a useful target list. "And don't bother me with that rival tribal crap. If they just offer someone up who isn't in their clan, and we figure that out, I will cut them off permanently."

Erin just laughed. "Nice bluff," she said, knowing that we had only a rudimentary grasp of Iraq's cultural intricacies. I certainly did not have the knowledge to level such a threat. But they didn't know that.

Next, Erin and I scheduled an emergency briefing with base leadership. But I had a different source of angst that I wanted to share.

"I'm afraid big changes in security might compromise the intelligence," I said. I wanted to remind leadership that any demonstrative show of force based on the information – an appropriate and somewhat conditioned response in the military – would attract attention. I wasn't suggesting that they not do it, only that they keep in mind what would happen when they did. Pacing the floor, I told Erin that significant shifts in security, a visible reaction, would tell the terrorists that someone had squealed. That risked exposing the source and would allow the attacker to shrink back into the shadows, where we would struggle to identify them. Knowing all of this, she simply nodded.

I returned to an earlier thought. "The whole thing could be a test of our security too," I said. I knew it was a stretch, but I wanted every possible scenario on the table. Erin looked up, half amused, half interested. A thought had come to me that Sunni insurgents might have leaked false information to see how Bucca security responded, and so the attackers could modify any future plans. If they watched everyone closely enough, they might even be able to identify the intelligence on base, as bearded civilians shuttled busily between known security offices. Erin reminded me that it already looked like an inside job, which gave some credence to my theory. I much preferred an escaped attacker and lives saved over the reverse, but it came down to our offering the best risk analysis possible.

★

"Sir, we have credible information from sources near Baghdad that the chow hall is a target for an attack," Erin began. "The report suggests the use of explosives, but we couldn't confirm the nature of the attack." The commander sat quietly. He had already received a call from his headquarters forewarning him about some critical intelligence, although he had no details. Calmly, he digested what he knew was coming. "The other problem," Erin continued, "is that we have no timeframe. Could be today, could be in two weeks. We simply don't know." She took more time to explain the apparent plot, as best we knew it.

Before she could finish, our counterparts in Army intelligence practically ran into the office with reports from Baghdad, detailing a similar threat, and the commander sat up straighter in his chair. After a long pause, he spoke. "Well, what do you think we should do?" Always a calm and calculating man, he looked at us, then at the Army major standing nearby. Erin went first. "Sir, the threat appears legitimate. And extra security around the chow hall would be best." The Army major jumped in: "We have the manpower to beef up checkpoints in and out, but only for a limited period. Otherwise, we have to rethink our perimeter security and patrols."

The lack of information on timing really put us in a bind. After letting both comments sink in, I spoke. "Sir, we think extra security around the hall for the next meal service would be best, and let it continue until we can flesh out more information from our guys up north." I didn't check with Erin before saying "we," but after working so closely for so long, we practically read each other's mind. "Hopefully, a sustained security presence will give the appearance of normalcy after a few days. Kind of like boiling a frog in water." I could see Erin grinning in my peripheral vision. "Extra security now and for at least two weeks would be best."

Erin then added, "Since it looks like an inside job, it could be a worker with access to a separate entrance. Food delivery, chow hall equipment, or something like that. Security should probably search every worker and all the food and equipment coming in and out." The major nodded in agreement.

Another long moment of silence, then the commander spoke again. "How should we increase security on the perimeter of the chow hall?" This was not our call. The Army major left the room and returned within minutes, with the security forces leadership in tow.

Tension hung in the air. They were advised on the threat in vague terms and recommended a series of upgrades, to include additional blast walls, armed guards, and random patdowns. We all knew that an attack would absolutely devastate perhaps the most strategic base in southern Iraq, at the crossroads of America's vital logistics routes, and home to the war's most notorious terrorists.

The commander waved his hand in agreement to all the recommendations. And with that, the security forces officers darted from the office and sent word to upgrade the security posture around the chow hall. The rest of us sat silently for a moment. Then, with his usual calm demeanor, the commander spoke up. "Well, I have to make some phone calls, so if you will excuse me." He swiveled his chair slightly and reached for the secure phone sitting on the corner of his large wooden desk.

Erin and I rose and quickly left the room. We didn't exchange words the entire walk back to the office. The chow hall would be open for lunch in about an hour. And I knew we both were contemplating whether a meal was worth it.

★

"Have our sources heard anything?" I asked our linguist, Abdul, soon after entering the office. He just shook his head without looking up and began punching numbers on his cell phone. He pressed it to his ear and mumbled, "No one has heard anything about an attack. I'm actually freaking some of them out."

"Please tell me you're not giving them details," I said quickly.

He waved off the comment and laughed. "Come on, Mr. David. They just hear attack and piss their pants. These Iraqis, they're not as tough as they pretend to be." He was from Egypt and felt the need to remind me that the pecking order in the Arab World had Egyptians at the top.

I noticed Erin typing furiously. "You all right over there?" She didn't say a word – a signal not to press her further until she was ready to talk. I sat down in my chair and stared blankly at the plywood partition separating our office from the neighbors, contemplating our next move.

We often had information unavailable to others, but this one had immediate consequences. Only a handful of people were fully aware that the entire base was potentially walking into a deathtrap. The old adage that ignorance is bliss had never rung truer. I briefly considered skipping lunch. My parents and sister would not hesitate to ask that I use my job to protect myself. They would never ask that I risk the lives of others unnecessarily, but they would ask that I avoid death if I knew it to be imminent. But if I used the information to my advantage, I wouldn't be able to live with myself. That would be the most consequential kind of insider trading.

Scolding myself for such a thought, I considered calling my parents. Our phone couldn't reach U.S.-based numbers without going through a switchboard. The operators would grant a fifteen-minute morale call to deployed personnel and connect the caller to a local number free of charge. And I had an advantage. The whole process didn't matter if the number I needed to reach was in the same area code as a nearby military installation. I could call the base directly, and the base operator would forward me to a local number. No one watched the clock on those calls.

"Well, MacDill Air Force Base is in Tampa," I mused aloud to Erin.

I reached for the tan, punch-dial phone and lifted the bulky receiver.

With the handset firmly situated between my head and neck, I pulled the cord so it did not tangle around the edge of the desk and used my free hand to dial the MacDill operator.

"MacDill, how may I direct your call?"

I knew the routine. "I'm trying to connect to a local number," I said.

"Okay, what's the number?"

I rattled off the phone listing of the home where I had lived for nineteen years. "Thank you. Please hold." And the line went dead for a moment.

Just as the ringing on the other end began, Erin blurted out from across the room, "Are you calling now? It's like three in the morning!" I froze. The phone had already rung twice. Without thinking, I slammed the receiver down with a thud. I had probably just startled my poor family – especially my father, the minister, who fielded calls in the wee hours only when something terrible had happened. I kicked myself for such a stupid mistake and vowed to call again at my first chance.

Erin spoke up. "Should we go?" The comment brought me back to our current predicament.

I said nothing.

"Forget it," she said. "We should be there."

"I agree," I replied without hesitation.

I turned to Abdul. "You coming?" He had been listening intently, and now he looked at me and back at Erin.

"I think I will stay here," he said. "I'm not interested in going." He tried to hide behind stoic body language, but I could see a slight look of humiliation cross his face. Abdul often sided with his pride, even to his own detriment.

"All right then, can I give you a letter to send to my mother?" I asked.

He shook his head vehemently and retorted, "Oh, you stop that right now, Mr. David."

★

Erin and I arrived at the main entrance to the chow hall to find a line of people backed up well beyond the typical staging area. Cement barriers extended several yards outside of the building, arranged so the line would snake outward from the main door about 25 yards. Several security officers frisked each person before allowing them to pass into the serpentine maze. The typical security review involved little more than an ID card check and clearing one's weapon.

Those who entered were met by more armed security forces posted along the pathway, rifles at the ready.

People were obviously taken aback by the sudden surge in security. Even the guards manning the checkpoint had no idea why, other than that an unspecified threat existed.

"Working hard, I see," I said to a security forces friend conducting patdowns.

"Yes sir," he said, looking up and grinning. His dark sunglasses hid his eyes. "You probably know more about this than I do, Mr. OSI." It occurred to me how hard it can be to read someone without seeing their eyes. I just smiled. "You think too much of me, kid. I'm the one in the dark." He just laughed and waved us through.

Erin and I zigzagged through the preconstructed walking path until we came to the stairs leading into the chow hall. A security forces sergeant manning the door raised his hand defiantly, signaling us to stop. "We let in five at a time," he said sternly. His gentle Southern drawl was hard to square with his commanding presence, shades, and rifle slung across his chest. The brim of his floppy, camouflaged hat sat snug on his head.

We stood there until three others came up from behind, nearly colliding with us. Without a word the sergeant swung wide the door, and the five of us marched up the stairs and into the hall.

The normalcy inside caught me by surprise. It was loud as always. The smells of cooked food wafted through the air. Loud and excited voices mingled in a massive chorus. Nothing seemed different. Few others even knew about the threat, and the armed guards and extra barricades no doubt sent a message. Still, many seemed to go about their meals without much regard for it. Life in a warzone.

I decided on comfort food for this meal and gingerly made my way to the grill line. Quickly I retrieved a ready-made cheeseburger and some sides, then made my way to a table on the left side near the rear of the building.

As usual, Erin chose to wait out the much slower line that collected where people decorated their salads with what I considered unnecessary precision. But the return of a salad bar after the base had gone without lettuce for several weeks was a treat for her and others. A few weeks prior, unknown assailants had hijacked the truck carrying lettuce as it passed through the border city of Safwan. The ringleader likely received consequences for that one. It was those hijackings that remained a thorn in the side of Air Force logisticians, and when they found out about our prohibitions from leaving the base to gather intelligence on those thefts, we were told, in no uncertain terms, that it would be fixed. I was skeptical, but even that small measure of support had given me hope.

Erin put the finishing touches on her big salad and circled back toward the tables. She scanned the room, standing there in her customary bright polo shirt and tan khakis. I was about to flag her down when I noticed that some British troops, apparently visiting the base for lunch, were now seated at the table directly in front of me. They had noticed Erin from afar and, like every other guy on base, were having a hard time ignoring her. She was undeniably attractive, and her civilian clothing amid a sea of uniforms only added to her allure. The six British soldiers began elbowing each other, pointing out the beautiful woman looking for a place to sit. I could hear the chatter: "Hey, why don't you ask her to sit here?" "Wave her down." "You're a bunch of ticks, you don't have it in you …"

The longer she stood there, the more excited the banter grew. Finally, one of the Brits waved at Erin and pointed to the empty seat next to him, which elicited some name calling from the others. Erin didn't see him. And I decided to have a little fun. I waited until they were fixated on her, then signaled for Erin from behind them. They had no idea the gentleman she was looking for was seated to their rear, and I hailed Erin at the same moment one of the British soldiers offered a second wave.

Erin acknowledged me with a bob of her head and began walking toward our table. The soldier thought she had signaled to him and panicked. His mates were in shock. The table was in an uproar of laughter and nervous chatter as she approached. "Is she seriously coming to sit with you? You're hideous." "She can't see that from over there." As the British soldiers ribbed each other, Erin slowly made her way down the aisle. Just as she came even with their table, she paused and placed her tray on the bench behind them, taking a seat in front of me.

The whole table burst into laughter. The Brits unleashed a torrent of ridicule on their mate, who flushed with embarrassment. "You really thought she was coming for you?" one said so loudly I thought Erin would hear. But she was oblivious to the whole encounter.

★

The moment of levity soothed my anxious spirit. In fact, I had momentarily forgotten the source of my restlessness.

Erin and I did our best to remain calm. We chatted about life and family back in the States, while intently scanning the crowd for anything and anyone suspicious. The security out front would prevent someone from entering the main door with explosives or heavy weapons. Such things would likely be secreted through a separate, less trafficked entrance. We watched the workers closely. We took a keen interest in any who seemed to linger near crowded areas, who wore bulky clothing, or who sported that famous five-mile stare common among suicide bombers before they detonate. Nothing looked out of the ordinary. And, within half an hour, Erin and I prepared to leave.

We returned to our office to little fanfare. The base hummed along as always. I felt relief, then frustration, knowing dinner awaited and who knew how many meals after that. The inability to pinpoint a time of attack grated on me. I fired up the computer to review some emails, only to find several from my father. My heart skipped. My earlier call would have populated a caller ID from a government number, probably MacDill, and they would undoubtedly have thought the worst.

My father's first email read, calmly, "David, if that was you, please call us back." I immediately grabbed the phone and executed the call process with the local installation. I explained the debacle to my shaken parents and apologized profusely. My dad had researched the number and, of course, found some online information that claimed such a number originated with chaplain services; another obscure link said it was mortuary affairs. They had redialed the number, only to reach a switchboard with no further information. Hasty emails to me went unanswered for hours since I was away from my desk. After a few words of fatherly reprimand, life returned to its usual nervous energy.

★

"Think we scared the bad guy off?" I asked Erin as we sat in the office reviewing intelligence reports.

"Who knows?" she said. "There hasn't been the kind of chatter you would think there would be when someone plans such a high-profile attack." It did seem rather quiet, despite the level of the operation. We hadn't received a follow-up from the Baghdad source. No new information seemed forthcoming. Then again, such massive attacks happened all the time without an inkling of news.

I continued mulling the inside job angle. The attackers would likely need to plant a device prior to a meal, if one was to be used. A bomber would need to be pre-positioned to avoid attracting the attention of roaming guards. Someone had to have universal access or be given access, especially now, after increases in security checks. That little piece of information might help narrow down the list of targets.

I tapped the corner of my desk with a pencil. Then a thought entered my mind and immediately blurted out of my mouth: "Where's Yani!?!" Perhaps it was his behavior weeks earlier that had caused me to think of him. The questions. His overall slimy demeanor. I looked over at Erin to gauge her reaction. Her eyes twinkled just a bit at the thought. "That's a good point." Before she could say another word, I called over to the base services office.

"He's gone. When does he get back?" I rattled off questions to the contractor on the other end of the line, then slammed the phone down in disgust. "Apparently, Yani hasn't been to work in days. He says that's not unusual, since he leaves to go home to Baghdad for several weeks at a time." Erin looked blankly at me, processing the situation. I didn't wait for a response. I turned to our linguist: "Abdul, call our guys close to Yani. See if they think he's capable."

★

Two weeks passed without an attack on the chow hall. And two weeks passed without our seeing Yani. The base commander decided to lift the extra security, after our contacts in Baghdad failed to confirm any more details surrounding the threat.

I ran into Yani almost three weeks later while crossing the basketball court at Bucca Square. He seemed a little more distant than usual. Less enthusiastic to see me. But mood swings were not unusual in him. Knowing this, I tried not to overanalyze his behavior, but I couldn't help scrutinizing his every word, shrug, and sideways look.

He claimed to have been taking care of his sick mother. I didn't believe it for a second. But I also had no evidence that he was involved. If nothing else, I used the moment as revenge for his inquisition of me weeks ago. "Yes, I know. Your mom is sick. It's all over the base," I said with obvious exaggeration. He looked at me confused, almost frightened, and stammered: "My mom, really?"

He just stood there in shock as I aggressively peppered him with questions. An accusatory tone cloaked in sympathy kept him off balance. "She's good. Umm. Fine. Why are you … umm …" Yani couldn't even complete his thought. I glanced at his feet, observing their position and angle. I watched to see if his hands came to his face at any point. Little tells that can often indicate deception.

I stood there, allowing him to stumble through a few more sentences with half-smiles and weak attempts at humor. At the first pause, I turned and walked away.

I saw Yani only once more during my remaining time at Bucca. We made eye contact, and that was it. Soon after that, the Braggart of Baghdad fell through the cracks and disappeared.

CHAPTER 18

Secret Squirrel

Erin hung up the phone and looked over at me. "They want to see us while they're here." I looked up from scanning my paperwork. It seemed that OSI command had petitioned enough on our behalf that an upcoming Air Force visit would now include a deskside briefing from us.

"Wow. All right," I said, unable to hide my surprise. It had finally come. The search for Zarqawi, the increased interest in our work, and our complaints about restrictions had produced a special visit to our office from Air Force command. We received word that a contingent of high-ranking officers would be arriving at Camp Bucca sometime in the following days and would make a special stop to see us. We had been through this sort of dog and pony show before, having provided security details to visiting dignitaries at Bucca. But this was different. We were to be a focal point of their visit, a detail conveyed to us privately from OSI headquarters. That meant there was added pressure to provide just the right balance of appropriate complaints and accompanying solutions. It would be our opportunity to put up or shut up.

★

Darkness fell over the base like a thick blanket, as it did every day. The black nights reminded me of late evenings in South Dakota, where I couldn't see twenty feet in front of me on drives through unlit mountain passes. It soothed a restless soul.

I was well into one of those few peaceful rests when what sounded like a massive party in the distance shook me awake. Sounds traveled well across the flat landscape of the base, so hearing far-off noises wasn't entirely abnormal. But the commotion had an unusual rhythm that seemed to be growing in intensity. I arose from bed more curious than alarmed and, just as I did, sirens began sounding inside the jail. I heard boots furiously crunching the gravel outside my hooch, as soldiers headed in the direction of the prison.

Propped on my elbow now, I fingered the flimsy plastic shades on my tiny window. Massive flames were spitting up from inside the prisoner compounds. I could faintly see guards sprinting back and forth outside the fence line. Floodlights rendered a brilliant scene against the black sky, as bright white light mixed with the orange glow of the growing flames. The scene was perfectly illuminated. Prisoners cheered as they danced at the edges of what looked like several giant bonfires. Inmates disappeared behind the growing blazes, only to reappear elsewhere, jumping and yelling. Others ran back to the hooches and reappeared near the fire, throwing objects into the flames.

Now I began to contemplate whether the prison could be overrun or, more likely, whether a handful of prisoners could use the moment to escape. A wholesale takeover was unlikely, but a distraction like this could allow a team of four or five inmates to sneak away. And even a small number of combatants let loose on the expansive base could pose serious problems. Slowly and methodically I prepared my rifle and wedged some magazines into the ammo pouches that lined the front of my vest. I decided to check before arming up with a t-shirt, workout shorts, and flip-flops. Swinging the door wide, I marched out to the front of my stairs and saw clearly that a prison riot had engulfed the entire internment facility.

I said a quick prayer for those night guards who were now trying to suppress a very violent situation. Having seen the homemade weapons of detainees, and the hate in the eyes of some of them, I was convinced we would lose an airman or two. There was simply no way those prisoners could be restrained, no way those fires could be extinguished, without losing an American.

An ATV flew past my room on the main road off to the right, dust rising as three passengers raced toward the scene. One sat in the back, looking as if he had been awakened from a deep sleep. He sat still, rifle strapped to his chest in crumpled fatigues that clearly had not been washed, his face registering a familiar mix of ambivalence and preparedness. That look of attentive resignation: ready to fight, even if resigned to fate.

The ATV disappeared into the blackness that extended just beyond the prison walls, beyond the reach of the floodlights. Startled free from my musings by voices, I spun around to see two soldiers who had climbed atop their hooch with two lawn chairs. Sodas in hand, these guys lounged comfortably while watching the riot in the distance.

Voices echoed from many yards away. I pushed back from my now reclined position against the hooch and peered around the corner of the trailer. A crowd had gathered. It seemed the racket had awakened the whole base. Hundreds of personnel lined the walkway between the trailers. Some sat in lounge chairs, others stood, observing the chaos engulfing the prison. I scanned the horizon to see that nearly every trailer had two or three people sitting on the roof, as if watching a drive-in movie.

I was reminded that the prison was well maintained. Security would keep these riotous inmates contained. I was glad that, in my vanity, I had not rushed outside in complete battle rattle.

Moments later I reemerged, to see the far-off guards standing at strategic corners along the upper walkway and lower levels throughout the compound. Even from my vantage, I could sense the menacing presence they exuded in their riot gear. No alarm in their posture. No fear in their faces. The guards stood stoically by as the prisoners ran about, screaming and shaking the fences. The guards' lack of panic spoke to their training. Inmates were simply ruining what few belongings they had, and the guards let them, like parents allowing a child to throw a temper tantrum until he wore himself out.

It had been only about fifteen minutes or so since I had climbed out of bed, and it dawned on me how hot it was despite the late hour. Sweat trickled down my face, and I wiped it away with my shirt.

Suddenly the crowd let out a massive cheer, and I raised my head toward the TIF. The guards had suddenly rushed the problem areas in a precise suppression. The planned incursion into the riot was swift and executed with overwhelming force. Before I knew it, the prisoners had been subdued and the fires nearly extinguished. The whole stabilizing operation seemed to take no more than five minutes. The prison fell silent, as lockdown and cleanup ensued.

The following morning, Erin and I could see that the inmates had also torched some of their own barracks, leaving them to roast under a hot sun. Later that day Joe told us that prisoners had somehow stacked mattresses and lit them on fire, creating the massive fireballs we had seen from the other side of the base. Only a small number of prisoners had been involved, but the fires made the chaos look more dramatic.

★

The days between notification of the big visit and the visit itself passed in the blink of an eye. Before we knew it, a one-star general and three colonels walked through the door of our modest double-wide office. The limited space forced a handful of staff to linger outside in the heat. The general immediately waved off the formalities. "Please don't get up, keep working. We just came to say thank you." He shook hands with our officemates. He spied the strange team in beards and civilian clothes. "You must be our OSI," he said with enthusiasm. "Yes sir, welcome," we said in unison. The colonels were on his heels, and we exchanged pleasantries just the same. And just as quickly as they had entered, the team of officers departed.

Erin accompanied the exiting entourage to the next office. Perhaps the meeting was to take place at a later time. Or maybe that's where Erin was headed. Before I could wonder further, I noticed one of the colonels lingering near the front door, speaking in hushed tones with an aide and looking as if he might reenter our office. I didn't want to retreat from decorum until I knew they had all left. I stayed in place, with all the formality I thought necessary. I was pleased I had, because moments later he did in fact reenter.

"Agent Grantham, was it?" I nodded, standing with my hands clasped behind my back at parade rest. Far too formal, to be sure, but training made it instinctive. "Let's talk about the situation here," he said quietly, and motioned me to relax and sit as he grabbed a vacant chair against the wall. His demeanor told me this was "off the record," a private conversation where I was at liberty to be as blunt as necessary.

This was not the way I had envisioned this briefing going. My notes were resting somewhere in a stack of paperwork on my desk, and I certainly wasn't going to shuffle through the disarray in front of him to find them. I briefly worried about appearing unprepared by rambling through talking points I had not committed to memory. Then again, this private meeting was clearly shaping up to be a short one. He leaned forward intently and glanced at his watch. I had his undivided attention, but only for a small window of time. I needed to capitalize on the moment by giving him a handful of facts, with examples and solutions. I reached back into my memory and grabbed three points, knowing that was what most audiences preferred.

I recounted the burdensome restrictions from an unyielding bureaucracy that, in my estimation, would not change its approach unless outside forces intervened. I reminded him that base leadership seemed privately agreeable to our requests but had no interest in fighting rules imposed on them from higher-ups. I also highlighted the lack of equipment – whether trucks, computers, or classified office space – although everyone in Iraq faced issues of resources to one degree or another. I concluded by saying that, whether it be material or permissions, we simply were not a priority to base leadership. "We are not worth the fight." I made every effort not to complain or sound whiny.

I then changed focus to reviewing the results the Air Force wanted from us, and how current predicaments impeded their need for intelligence. "We could be of real value with some support," I said, trying to end on an upbeat tone. Although my skepticism that anything might change had nearly emptied my well of enthusiasm, I wanted to show eagerness at the possibilities if only our office were given the chance. He listened intently throughout, acknowledged what we had accomplished with minimal permissions, and didn't hide his displeasure about the restrictions. More than once, he shook his head in disgust. His demeanor said that this was no exploratory conversation; I had confirmed for him what Air Force leadership feared. "We are hearing this from multiple places," he said. "And we will get this fixed." A look of determination crossed his face. It was a moment of openness that most young officers do not get to experience from such a high-ranking officer. I appreciated him for it. There was something empowering about a colonel speaking bluntly to a young officer like me. It could have been a show, his way of building rapport. But I didn't get that sense. He knew the liabilities in displays of unguardedness, and he did it anyway.

Our meeting lasted about fifteen minutes. The colonel and I shook hands, and I escorted him to the next building where his entourage awaited. I returned to my desk and fired up my computer. Our office in Kuwait wanted to know the outcome of the visit.

For the first time since I had arrived at Camp Bucca, I felt optimistic. I sat, feeling relieved and a little inspired. Not for myself. We would be long gone before any changes materialized. My excitement was for those who would fill our spots when we redeployed in a few months. Erin and I would leave a growing network of sources, the first of its kind at Bucca, and possibly with new permissions that would result in even greater information.

Before corresponding with our main office, I scanned through unread messages. I saw an urgent notification from the same guy who had previously disallowed "arousal cans." I laughed to myself, hoping this email would be something similar. "A sandstorm will be hitting the base this evening," it began. "Please store all your gear and cover any important electronics." Reading on, I saw that he didn't disappoint. "Due to the *impalement* weather, we will be canceling all outdoor training activities for this evening."

Erin and the others arrived back just as I burst into laughter. "He did it again," I said between breaths. The whole trailer was soon laughing hysterically. We could hear the commotion next door, as people yelled through their opened windows to passersby, warning them of spears falling from the sky.

★

Whether it was just dumb luck or new energy that led to better results, a high-profile source visited the camp not long after our fruitful meeting with Air Force command, to outline how Shia militia were helping move war materials straight up the corridor linking Umm Qasr and Basra. He drew a map of the routes; he had times of transfer and vehicle descriptions. Another source relayed information regarding Al Qaeda movement near the border with Iran – a curious development that would pique the interest of leadership, since Iran and the Sunni insurgency were not especially close. The detailed information improved our intelligence on Iranian influence in the south, which provided a clearer picture of regional geopolitics for policymakers.

A couple days after that, an informant returned to Bucca from his weekly visits to Baghdad with incredible information on Al Qaeda terror squads, detailing how Zarqawi's tactics had escalated to an almost incomprehensible level of bloodletting. The roving bands had moved to open slaughter and random murder.

We had an invaluable resource in workers traveling to and from Baghdad. They shared information with us, which they feared supplying to anyone around the capital city. Distance gave them comfort. But there was a greater problem revealed in all this. The local labor force the U.S. military relied on was under threat.

Soon after that meeting, I drafted one report highlighting how workers at Bucca were increasingly fearful of being identified and resorted to actions that, if they became more widespread, might hurt sustainment. Some workers tried desperately to find places to live for extended periods near Camp Bucca, so they would not have to go home to Baghdad. Others simply felt the risk was too great and never returned to work. Still others went to work elsewhere.

The threats and hit squads were nothing new. But the consequences of Al Qaeda's ratcheted violence had changed the often overlooked dynamic between support services and local labor. Iraqis were voluntarily displacing themselves. The workforce appeared to be growing inconsistent. The shifting populations and constant turnover were emerging, in our estimation, as the next security hazard. Maintaining the infrastructure of the base would quickly become a problem, if Bucca could not preserve a trustworthy body of local workers. Perhaps the violence was ramping up because we were closing in on the Al Qaeda leader.

★

Not long after that, reports trickled in that U.S. forces had hit a safe house in Baghdad, missing Zarqawi by only a few hours. The timing reinforced my belief that American forces were closing in. Al Qaeda in Iraq had released a propaganda video featuring its Jordanian leader firing an automatic weapon set to music, in an undeniably impressive advertisement. In his customary black clothing, with a black scarf covering his head, Zarqawi projected intimidation.

As it happened, the raid on that safe house uncovered, among other things, the outtakes from that propaganda video. The original taping showed Zarqawi standing amid a circle of helpers, looking confused and unsure of himself. The fearless Al Qaeda leader looked inept at operating the weapon he was given. At one point, an underling had to jump in and unjam the rifle for him. And in a fitting conclusion, the gun was passed off to another underling, who naively grabbed the now extremely hot barrel barehanded and recoiled instantly. I recalled how Makeel, the Al Qaeda prisoner, had claimed that Zarqawi was a dunce.

The U.S government renewed my faith in our ability to counter-message by releasing the uncut version, so Iraqis and the world could see the real Zarqawi. The raw footage quickly went viral. We all felt that an embarrassed Al Qaeda was a significant victory for us, since their recruitment relies heavily on a carefully manufactured reputation. "We need to send this to Joe," I said to Erin. "Remember that his guy said Zarqawi was an idiot. He might get a kick out of this." A message to Joe was met by an immediate response. He said they had planned to use the film to persuade other prisoners that Zarqawi was a fraud. I smiled at the thought.

★

A loud ring from Abdul's cell phone, unattended on the desk nearby, broke the unusual silence in the office. Our Egyptian linguist darted into our office from the adjoining room and grabbed for the phone. He hung up and stared intently at me. "We have a visitor at the gate. We must go now," he said.

A light breeze, complemented by a calm mid-afternoon sun, made our long trek to the main guard shack at the front entrance a bit more bearable. The plywood hut, about the size of a single wide trailer, sat 50 yards inside the base perimeter. It offered limited space for source interaction but enough concealment to conduct brief meetings out of sight of those ever-present prying eyes beyond the walls of Bucca.

The smell of diesel stung my nostrils as we approached. I could see what always looked like an endless column of delivery trucks awaiting inspection. Small plumes of dust kicked up where the exhaust from idling trucks bounced off the sandy desert floor. The long line of haulers sat at a standstill near the entrance; the line ended somewhere around the corner and off base, out of sight. I spied several drivers seated under the shade of a small patch of trees just beyond the main road, as security methodically securitized every compartment of their semis. As we passed, one officer crawled on his back with some sort of inspection tool to view the undercarriage of a truck. The incredible amount of time it took to process one hauler made me wonder how we managed to receive supplies in any kind of timely manner. And yet the whole thing was absolutely necessary.

Months earlier, a story had circulated that inspectors had found bomb-making materials in a semi truck delivering supplies to Bucca. The keen eye of a young airman uncovered a case of ball bearings stuck deep beneath the passenger seat. Although harmless by themselves, ball bearings were a primary component in suicide vests and improvised explosives. Add to that the effort made to hide them behind a seat, and that they were completely unrelated to the shipment, and they struck the young airman as suspicious. The discovery led to an investigation, which found additional connections between the driver and known or suspected terrorist organizations. Supposedly, an extraordinarily successful military operation ensued.

Even still, the long security process didn't need stories like that to prove its value. The ramifications of poor oversight were obvious. Few complained, not even the drivers who cruised through a warzone as slow-moving targets of opportunity and then waited for hours to be inspected. (At the same time, it's true that they were well compensated.)

As we arrived, guards at the main shack directed us to the front gate. Apparently, locals had not been granted access to the second layer of security. I paused before proceeding, knowing that this presented a dilemma. The front gate shack was barely inside the base walls, requiring us to be fully geared up. Erin and I had anticipated that and come prepared. But the real problem involved the shack's lack of space and position clearly visible from the road. We certainly did not want to compromise the security of a willing source. Nor did we want to expose ourselves to possible lookouts monitoring the foot traffic at the entrance. I knew then that we would have to provide the illusion of secrecy. In other words, we had to make the source feel we had made every effort to make the meeting as clandestine as possible, knowing full well that there was little we could actually do to conceal our work. At least the appearance of stealth would hopefully establish a rapport with the source before a word was spoken.

We took a long, winding route to the back entrance of the shack, a walk that was largely concealed by cement barriers and Hesco bastions. We maneuvered in and out of the serpentine layout of these barriers near the front gate, to ensure that anyone watching from the road would have a difficult time seeing us. We advanced toward the smaller outpost and snuck in through the rear door.

The booth had an opening that resembled a small wet bar with a serving counter and bench. Three or four other Iraqis lingered about, probably waiting on paperwork. I could clearly see an old Iraqi man seated on a bench, just beyond the gate that divided the U.S. base from Iraqi territory. I thought I recognized him as someone we had met during our routine security patrols along the highways. I told the guards to call him over casually enough so as not to alert the others.

The unassuming elderly gentleman slowly made his way to the bar. We huddled near a corner, so people could see him but not me. I had briefly considered having him brought behind the shack so we both could be concealed, but that would have risked arousing curiosity. Instead, we made it look like a routine encounter, a source meeting hidden in plain sight.

The man had agreed to contact us should he see or hear of any threats to the base. And he didn't disappoint. With an almost unconcerned demeanor, he explained how he had witnessed strange activity near one of the checkpoints on the main road leading to the base. He said that young men unfamiliar to him had gathered along the roadway several times, appearing to do nothing but loiter. Born and raised in the immediate area around Bucca, the man knew his surroundings and was certain that they were up to no good.

He waved off any mention of compensation, before I had a chance to mention that we had none to give. Our other, less appealing offers were met with similar lack of interest. As best I could tell, he considered himself a concerned citizen who simply wanted to extend a small measure of support to preempt any violence. It seemed to me that he valued the American presence, but was more concerned with stopping hostilities in his neighborhood. The intelligence led to some aggressive counter surveillance, which presumably scared off what I believed to be operatives, monitoring incoming and outgoing American convoys. It turned out that random engagement with locals on the street proved useful.

Before we left the shack, two young security forces airmen stopped us. They were no older than twenty, and both had big grins plastered across their faces. "We had quite a time with him earlier," one of them said. "He was trying to give us the code word, but he sounded like he was chewing on his tongue." They both started laughing hysterically. I stared at them, puzzled. Then it hit me that we had given instructions to ask for "squirrel" – a reference to our moniker "secret squirrels." It was the only thing I could think of at the moment, a word I knew would not be confused with anyone else on base. But we had also inadvertently provided one of the more difficult English words for an Arab to pronounce. The old man had apparently tried to say "squirrel" for about five minutes, before the guards finally recognized the word. "He said 'ski-wall,' 'sicque-wheel,'" the other guard said. "You guys are cruel."

CHAPTER 19

Poor Sheikh

It was graduation day and almost June. The long, cumbersome process toward handing over Camp Bucca prison to the Iraqis training as detention officers was about to take one more step. Erin and I graciously accepted the invitation from the Army instructors of the academy, who had quietly helped us identify sources within the school who reported on local issues for us. The Army National Guard from Rhode Island were among the most professional operators at the camp. I marveled at their effort, and at the patience they exhibited when training young Iraqi men who had few relatable skills and little interest in doing the work.

As we made our way to the ceremony across base, I mulled over how a majority Shia security force would soon oversee an internment facility full of incarcerated Sunnis. The American government had every intention of transferring authority to the Iraqi government as soon as possible, regardless of the impending confrontation between the rival Muslim sects. It was no doubt a welcome relief to allow someone else to manage the sensitive location and its unyielding political drama, particularly the patently false accusations of abused prisoners routinely provided to human rights observers, themselves all too interested in that information, which would disparage America's reputation. This pending handover already felt like a pyrrhic victory.

That war-induced bitterness slowly welled up inside me. And I hated that. A fatalist mindset, a defeatist attitude, has no place in war. I reminded myself: "God grant me the serenity to accept the things I cannot change, the courage to change the things I can, and the wisdom to know the difference."

We arrived to see five or six rows of folding chairs lined neatly along one side of a dusty parade ground. The gleaming white seats and some ribbons and other décor sprinkled on tables, all set against a backdrop of military tents and vehicles, gave the space a touch of formality. Base leadership took the front row as we found vacant seats toward the rear. I was impressed with the labor American forces went through to ensure this graduation had the trappings of any official military occasion, despite the woeful lack of material.

The ceremony began on time, undoubtedly due to American management. Iraqis were notorious for treating appointed times as mere suggestions. The Iraqi guards launched into what was supposed to be a formal military parade. Instead, it resembled a sort of organized chaos, complete with a steady stream of unsynchronized and incomplete movements. Some enthusiastic students marched completely out of step. Others hardly attempted the performance, acting supremely uninterested in the whole process. Still others looked confused, uncertain what to do next. For their part, the Iraqi instructors of the camp looked professional and polished. The whole thing lasted about ten minutes and ended with a hearty clap from the onlookers.

The Iraqi guards gathered to one side of the parade ground, as the commander took center stage to offer a few words of gratitude and encouragement. The soon-to-be graduates stood with a noticeable collective nervous energy. It wasn't long before the lead American trainer announced to the crowd that the graduation was official. Finally able to release their emotions, the trainees began jumping and shouting with excitement. The previously uncoordinated and uninterested guards joined together to form half-circles and, with arms draped across each other's shoulders, began rotating in unison, singing and kicking their legs back and forth. The famous *Dabke* dance of the Arab World had broken out. Apparently they did have coordination after all. The thrill seemed genuine and was encouraging to watch.

After a few minutes of coordinated jubilation, the circle collapsed on itself as men ran into the center, jumping and chanting in Arabic. Their enthusiasm made me feel, if only for the moment, that the handover might succeed. Their joy gave me hope that Iraq could survive. While I remained lost in thought, my linguist leaned in and shouted over the noise. "That chant they are yelling…" I nodded without making eye contact, indicating for him to continue. He leaned in closer and said, more softly, "They are chanting for Muqtada al-Sadr." At this, I turned my head and pulled my glasses down from over my eyes. "Are you serious?" He nodded, now more sheepishly, as if upset with himself for ruining a hopeful moment.

I couldn't believe it. They were cheering the infamous Shia militia leader who had fought American forces and aligned closely with Iranian saboteurs. Whether they realized it or not, they were aligning themselves with an anti-American radical and announcing it at the graduation of an expensive training paid for by the United States. If their chants were any indication of their loyalties, the nefarious Shia leadership and the Iranians already had a foothold.

Looking back, this hyper-Shia atmosphere at the celebration partly explained the Sunni discontent that ISIS would thrive on years later.

★

The bad taste in my mouth from the graduation remained, as May turned to June. The Sunni-Shia fighting continued. Iranian encroachment continued to seep into Iraq like a deadly, invisible cloud, a sort of wartime carbon monoxide, slowly covering more and more territory and leaving death and instability in its wake. Iranian infiltration butted up against the hyper-Sunni expansionism of Al Qaeda in Iraq, escalating tensions to a point that the United States could not manage. Or so it seemed from my vantage point in the south.

As I considered the conditions and how I might articulate them to my replacement, who would be here in a month or two, a high-value Al Qaeda prisoner named Abu Alwai sat in a jail cell at Balad Air Base in Baghdad. Alwai had recently given up some critical information on Zarqawi to U.S. interrogators who, after growing frustrated that favorable treatment had failed to yield the information they knew he had, threatened to move him to the general prison population. The prospect of losing favored treatment, such as access to special food and American films, loosened Alwai's tongue.

As Erin and I slogged through the day's intelligence on June 7, 2006, two U.S. Air Force F-16s positioned themselves overhead of Hibhib, a predominantly Sunni village approximately eight kilometers northwest of Baquba, a suburb of Baghdad. The quaint town resembled a farming community, with a canal, a two-lane road, and acres of palm groves. One of the pilots confirmed the coordinates of a target with headquarters. A house set off a dirt road near a large patch of groves sat undisturbed. A moment later, a massive explosion rocked the town. Sand and debris spewed in all directions, as the two 500-pound, laser-guided missiles hit their mark. The perfectly executed airstrike demolished the home, killing everyone inside.

★

I was preparing to leave the office. We had a long-planned source meet we hoped would lead to precise information on Iranian corruption within the Iraqi state police. I did one last review of the data I needed to sound knowledgeable about the division of labor within Iraqi law enforcement. The chair I sat on had seen better days. Broken in the back and now listing to one side, the once high-end office chair could hardly hold my body weight, much less the armor, ammo, and weapons I carried. Just before I rose from it, a message populated: "Got him." I scrolled a little further down the classified message and saw a two-line statement, notifying all intelligence units that Zarqawi had been killed in an airstrike north of Baghdad.

I sat still for a moment. An undeniable feeling of accomplishment washed over me. It was not a proud, boastful feeling, per se. It felt more as though a major task had been accomplished. I thought back on all the work we had put into disrupting Zarqawi's network, as small in scale as our work had been. I considered all the work U.S. forces had done fighting Al Qaeda in Iraq, all those who had died.

I learned later that Abu Alwai had revealed how Zarqawi routinely met with a spiritual advisor named Sheikh Abd al Rahman. The prisoner had revealed that the spiritual leader would physically move his family from one location to another, just prior to meeting with the elusive leader of Al Qaeda in Iraq. Once American intelligence located Rahman, getting Zarqawi was only a matter of time.

I canceled our meeting, threw off my gear, and phoned headquarters. They had little else to share besides what I had read. I could feel the contagious enthusiasm of the moment through the phone. I then excitedly called Joe in the prison. "Do his comrades know yet?" I was referring to the Al Qaeda higher-ups in the TIF. I could practically hear the smile forming on Joe's face. "Well, word is getting around," he said. "A few think it's just us infidels trying to mess with them. Soon enough, though." Before we hung up, Joe said, "I'm going to talk to Makeel."

Joe told us over lunch later that day that Makeel had entered the room in a state of feigned sadness. "Oh, Mr. Joe, what a sad day it is today," the Al Qaeda operative said. Makeel hung his head low, concealing his face. "Oh, my heart breaks. My poor Sheikh is dead." He put his face in his hands and let his shoulders convulse, as if crying. Joe said he sat there with a straight face until Makeel started laughing, face still hidden in the palms of his hands, and finally lifted his head to reveal a big grin.

Joe was a bit caught off guard. "Aren't you disappointed?" he asked, genuinely curious.

Makeel waved the question away with one hand, while he pulled up a chair with the other. "Whatever. You think I care about him? Good riddance. I could care less."

Joe said Makeel laughed about the whole thing. "I'm surprised he lasted this long," he said.

Zarqawi was dead. And I was going home soon. The highs and lows that war had introduced, the giant swings in momentum and fatigue, excitement and disappointment, would be coming to an end. But Bucca would make sure that the emotional rollercoaster continued until the last minute.

CHAPTER 20

Beehive

The main office phone rang just as I entered. I watched out of the corner of my eye, as Erin reached for the phone. "OSI, agent Berlow speaking." I began rummaging around my desk, which caused Erin to wave her hand in frustration for me to be silent. I turned around; her face was tense, and she put her palm to her forehead in disbelief. Something bad had happened.

I could hear someone frantically relaying information to her on the other end, but I couldn't make out the details. Erin wedged the phone between her shoulder and her head and pointed at me. I began to put on my gear, but she waved me off. "Your note pad and camera," she whispered.

I dropped my gear in confusion. She hung up and looked at me, pained.

"Some airman shot another airman in their hooch," she said. "That was Army CID, and they're begging us for help since it's Air Force on Air Force."

We were intended only for counterintelligence work, and here we were faced with yet another criminal situation. But Erin and I knew there were times when the spirit of the mission trumped the letter. Every available resource would be needed, and we had to move quickly to secure the crime scene and gather witness statements. "Beg for forgiveness rather than ask for permission," I said as we bolted for the door.

We arrived at the CID office after a dead sprint, to find the small trailer swarming with people. Like a beehive after a violent disturbance, the place was in frenzy. We ducked inside to find a similar commotion.

The doublewide trailer had a lobby large enough for three or four standing adults. Six or eight small offices lined the walls and were separated by a thin walkway. Commanders barked orders down the hall, at no one in particular. Phones rang off the hook from every corner of the building.

It seemed like any available floor space was taken by the anguished and harried.

One young woman sat curled up in a ball, eyes bloodshot from crying, still in her workout gear. Other Air Force members lined the hall, some panicked, some already overcome with shock, their eyes vacant. Witnesses were shuttled hurriedly to and from small offices. Others crammed in a corner, recounting what they had observed to rushed investigators.

Agent Smith turned a corner just as we began to push our way down the hall. She immediately turned on her heels and escorted us to the commander's office in the rear of the trailer. Every office we passed was the same scene: panicked, hysterical.

"His face was white as a ghost when they put him on the stretcher," Smith said over her shoulder. "We might be looking at a homicide." With that, she moved past us and ran back down the hall.

"So what the heck do you want to do?" the CID commander said, swiveling in his chair. He set down the phone receiver and propped himself up on the desk by his elbows. His hands came together to form a steeple, a subtle tell that someone felt confident.

"Well, what the heck happened, exactly?" Erin retorted.

The CID commander briefly explained that an airman had fired his handgun while in his room, and the bullet had struck another guy who was apparently lying on his bed or standing along the wall. No one was sure. "The shooter claims he was cleaning the weapon."

We rolled our eyes in unison. "That's bull," Erin muttered. We all knew that was the excuse someone gave after carelessly handling a weapon. All of us agents carried loaded sidearms always, and we knew it was virtually impossible to clean a loaded gun, and more unlikely that it would discharge by accident.

"We're not buying it, but he's not cooperating," said the commander. "I need to know if you want this case. Because if you refuse it, we will take it. And it will not be a joint case – it will be ours." The commander clearly still harbored some resentment over our refusal to take the stolen truck case. He was posturing now. He assumed, wrongly, that the involvement of Air Force members would spur us to demand jurisdiction, and he was prepared to counter. But his bravado proved unnecessary. Erin and I both agreed that the Army should take the case and that we would assist wherever possible. "Even if we wanted to, we don't have the criminal resources," Erin said. We would labor at articulating our mission to the very last minute, I realized.

The commander was now less defensive, and a bit taken aback that no debate had ensued. Momentarily caught off guard, he busily shuffled some things on his desk in an apparent attempt to gather his thoughts.

"Well," he began, eyes down, still moving papers around his desk. "First I need someone to visit the victim, if he is still alive." He looked up now, more certain. "Get testimony, photographs, and anything else that might be pertinent to the case."

"I got it," I told Erin. I knew she would have to be the one to start making phone calls to our bosses in Kuwait. It would be a long night for her.

I opened the door to the closed office. The scene was just as manic as when we had arrived. Pushing through the crowd on my way to the front door, I overheard a witness describe the victim as lifeless. The airman had been swiftly delivered to the trauma unit only ten minutes earlier. That wasn't a good sign.

I was able to snag an ATV parked outside and wormed my way through the base. Two minutes later, I skidded to a stop in front of the hospital, ripped the key from the ignition, and pushed through the first set of doors, not even considering if someone was on the other side. I made haste to the front desk.

I had not anticipated its cleanliness, even though it was the base hospital. The clinic looked surprisingly clean, even sterile – an unusual sight in Iraq. It resembled a fully functioning small medical facility that you might see in the States.

"Airman was brought here, gunshot wound." I flashed my badge and kept words to a minimum.

Without missing a beat, the desk clerk phoned some office, presumably in the back, listened for a second, then slammed the phone back down. "Follow me," she said quickly, and we darted through double doors to my left. Without a word, the clerk pointed to a second set of double doors about five feet ahead. I rushed toward them and pushed them open, to find a doctor standing there.

Back home I had run several death investigations, and body language told you a lot in those first few minutes after arriving at a crime scene. The doctor's look was a familiar one. I stopped abruptly, almost out of instinct, as my heart sank. I knew the kid was dead.

The possibility of getting details from the victim was over. Now it was a death investigation, which required that I get information from the doctor and see the body. I gathered myself, slowed my breathing. No need to rush my portion of the process.

We still had not said a word as the doctor ushered me around a corner and through a door to a medical prep area. He tossed me a gown and gloves from a tray sitting on a ledge, then escorted me into an operating room. The white walls were hard to look at: the light reflecting off them was almost blinding. The scent of cleaning supplies mingled in the air with the lingering stench of sweat and blood. I moved toward the body, still limp on the operating table. The freshly sterilized floor squeaked noisily beneath my feet.

"We tried," the head doctor said, dejected and exhausted. "He was shot through and through." He pointed to the exit wound under the victim's armpit. The bullet had gone through his chest and hit the sac around his heart, causing uncontrolled bleeding. "There was literally nothing we could do."

I tried to calm my nerves. There was no need to rush. In fact, I now needed to be methodical. I had to exercise enormous patience with everything I did, lest I miss an important detail that could be used as evidence in criminal proceedings. These types of dramatic swings from full speed to thoughtful precision were great for conditioning one's body for life.

I reached for the camera dangling from my neck and steadied my hand. For the next twenty minutes, I photographed the body from every angle. This was a routine procedure that I had conducted more than once before. No one said a word. The "click … click … click" of the Nikon camera echoed off the pristine white walls. It was the only sound in the room.

"What happened?" the doctor finally asked, breaking the silence.

"Don't know, but sounds like someone was playing with their gun and this kid paid the price."

"Ridiculous." He swatted at the air. "What idiot would play…" His voice trailed off. I wrapped up the photography and took a statement from the doctor. I could expect the medical report within the next few hours, he said.

I dreaded what came next. I would notify CID that he was dead, and that news would only worsen the atmosphere in an already grief-stricken office. More than once in my short career, I had found myself having to deliver devastating news to unsuspecting loved ones, then having to continue through the investigation despite their grieving. That would be the situation as I returned to the office. Those difficult moments came with the job, but they didn't get any easier.

We identified the victim as Airman 1st Class Carl Jerome Ware of Delaware. He had been in the Air Force for about two years, a security forces specialist out of Hickam Air Force Base Hawaii. I learned later that he left behind a pregnant wife, Christine, and a one-year-old daughter named Caitlyn. His second daughter was born six months later. She was named Carly, after her father.

After months of lying about cleaning his weapon, Airman 1st Class Kyle J. Dalton admitted in court that he had pointed what he thought was an unloaded gun at Ware and pulled the trigger. To my mind still a questionable admission, if not an outright lie. In November 2007, almost a year and a half after the incident, Dalton pled guilty to involuntary manslaughter and the unlawful aiming of a firearm when deadly force was unnecessary.

CHAPTER 21

Processes

I casually thumbed the corners of the paper in my hand. Both exhausted and relieved, I let out a long sigh. The office was empty. The sound of a humming computer filled the dead air. A feeling of accomplishment accompanied the sinking realization that this deployment would be coming to an end. I sank deeper into my faltering chair and allowed the finality of it all to wash over me. The archaic text and sterile language, the boring, formulaic layout consistent with government paperwork, all belied its importance. We had received our orders to return to the Kuwait office for out processing. This single sheet of paper was my ticket home. It dawned on me how many others would never experience this moment.

I briefly considered how those killed in action would never experience the joy of seeing loved ones on their return home. Handshakes from fellow veterans, pats on the back from strangers. Moments at the kitchen table, sharing stories with friends and family: these things were left up to the rest of us. I recalled those moments in war movies when a dying soldier pleads with a comrade to tell his story. Those had always felt contrived, dramatized for the silver screen. And they were. But at their heart was the bitter reality of war: that the dead cannot tell their own stories.

It had been a week or so since the deadly shooting. Army CID had taken control of the case, and we served as advisors as needed. In truth, Erin and I had virtually been cut out of the loop on that and other intelligence work. Two fellow agents who had arrived a few months earlier began taking over office affairs. This handover was one of several processes that represented the coming end of a deployment. And those so-called "processes" involved more than the humdrum, administrative tasks that the word implied.

★

The office was empty. The only sound came from a light wind that rattled some loose metal that dangled limply somewhere outside the window. I reviewed my checklist surrounded by an unusual quiet. Each process I needed to complete before leaving the warzone came with its own emotional, physical, and intellectual challenges. I had to begin with getting a doctor's clean bill of health for the laborious travel home. The IT guys would soon discontinue my access. Finance now had a copy of my orders and would require that I settle any outstanding debts. The list went on and on. Completing out-processing obviously took time away from my other work. Add to that how I now began stuffing multiple bags full of gear, clothing, and travel items for the trip home, and it became clear I would be doing little to support the war in these final days.

Inevitable arguments soon followed with multiple people responsible for the many legs in my travel schedule. It was no easy task to get a bus scheduled from one person, a plane ride from another, and a commercial plane to the United States from yet another. I didn't have a unit to redeploy with, so it was up to me. I joked to Erin that the Pentagon will move heaven and earth to get us into the warzone, and we had to move heaven and earth to get out.

It was a curious place to be mentally. I had begun to modify my behaviors that were needed only for the warzone, while maintaining them to a degree, because I hadn't yet left. Preparing to leave brought on a kaleidoscope of feelings, an unsettling swirl of emotions and actions, that would take months to settle after my return to the States.

Shuffling through papers, I thought of Kuwait and the posh hotels that awaited me there. I thought of my family, my church, even

South Dakota, a place I had grown fond of in my time away. I was reminded of the comforts of home every time I jotted down my U.S. address on outgoing mail. And while that process made me long for familiar surroundings, there was a part of me that never wanted to leave Iraq. The camaraderie, the risk, the rawness of it all, fed that inner animal spirit that craved adventure and conflict.

It became clearer to me how so many veterans of every war had failed to make the transition from conflict to civilian life. That kaleidoscope was a never-ending and unnerving parade of brilliant and maddening illuminations that left them disoriented. It was no wonder that some felt convinced that taking their own life was the only way to escape.

All the activities related to leaving helped us naturally and gradually disengage from operations. Erin and I knew it was inadvisable for us to be involved in the cumbersome exit process, while also playing prominent roles in high-profile investigations or operations. We would punch that timecard if an operation called for it. But for the most part, a case or operation should no longer depend solely on us. Indeed, once our feet hit U.S. soil, we would take up cases left dormant in our absence. Fraud, narcotics, counterterrorism – we had no choice but to switch out of warzone mode and back into credentialed federal agent mode. Case folders, witness statements, executing search warrants, it all began in earnest soon after settling back in. I eagerly anticipated that transition.

★

I strolled between trailers, collecting paperwork with a nagging feeling that I might at any moment be on the receiving end of a mortar round. Erin and I were under no illusions that the gods of war would allow for any mental or emotional transition, despite our preoccupation with packing and out-processing. A small part of me clung to the naive hope that we could pack our bags and enjoy a last few meals with comrades without major responsibility. But the bad guys didn't check us off their target list just because we were headed home.

My return to the office threw me back into the war, as intelligence continued to flow in. Informants and outside reports recalled how Al Qaeda in Iraq continued to kill with a brutality that seemed to have no strategic value. The bloodletting was just that. A part of my mind was on the war, and a part on home. I labored in my efforts to offer analysis on what had become a different set of complicated circumstances, with Zarqawi now dead.

Al Qaeda had responded to Zarqawi's death with violence as we had expected, all under the new management of Abu Ayyub al-Masri, the mastermind of the ambush and torturer of the two U.S soldiers we had spent days trying to find. But that single airstrike on a little house in Hibhib proved a debilitating blow to Al Qaeda in Iraq, leading to its precipitous decline. Masri would essentially manage Al Qaeda's slow collapse, until a former Camp Bucca prisoner named Abu Omar al-Baghdadi resurrected the group and rebranded it into the fearsome Islamic State.

In the front of my mind was the attempted transition from American to Shia leadership over the prison, which had far greater political and economic consequences now as the date neared. The risk to the U.S. military hinged on a smooth handover. But the U.S. government calculated the risk to be worth it, in the context of Bucca's overall operation. It was an understandable decision.

But an easy changeover didn't equate to a trustworthy Shia leadership. We knew instinctively that Iran would work to coopt many of them, if they hadn't already. Those Shia Muslims in the south would potentially occupy Iraq in our wake. That meant doom potentially awaited this country, when Sunni militancy and historic discontent fell under the sway of Shia leadership and Iran's meddling.

★

I began organizing debriefings with our sources amid the out-processing. I contemplated the reactions that our departure would illicit. Erin and I had done our best to begin transitioning all our source work to other agents. But we also knew that, despite preparation, some sources whose relationships with us were based intangibly on trust, unquantifiable and forged in the fires of war, would make the handover process challenging.

We went through incredible training to understand the process, because it was so important and so delicate. Experience and training taught me that some sources were more dedicated to the mission than to any one person. Others were less inclined to accept a replacement. Add to that a loyalty to our linguist, who earned a different level of trust with sources, and one could see how delicate this could be.

We met first with our most important informants. I thanked them for their work and spoke glowingly of the new guy, whether he was present or not. How I honestly felt about a fellow agent never mattered here. It was our responsibility to make him or her out to be the coming Messiah. There was admittedly some comfort in not hearing from the source again, should that agent not measure up to the endorsement.

One source, an older gentleman who had seen his country go through all kinds of conflict – from the Iran-Iraq War of the 1980s to the Gulf War and the invasion of Kuwait, and now Operation Iraqi Freedom and the latest Sunni-Shia civil war – met with me a few days before our scheduled departure. We convened at a wood picnic table on the outskirts of Bucca Square, during the heat of the day, when most people were either working or asleep. The area was practically deserted, just as I had hoped. Jamal was accompanied by two subordinates. I pulled some ice-cold Rani drinks from a plastic bag as we all sat down. These fruit sodas, popular in the Middle East, had chunks of fruit swimming in them. I couldn't get enough of them. And on this hot day, my linguist and our guests were just as glad as I was to enjoy one.

I wasted no time. "I'm leaving in a few days. You knew this would come eventually, so I hope you will be willing to work with our new guy. He's experienced, and better at this than me." The linguist interpreted those comments with relative ease. The words registered with Jamal a few seconds later, and he grinned ever so slightly. "Mr. David, you are very good at this," he replied. "I will be the judge of the next guy." This Iraqi, old enough to be my father, knew the system, knew the source-handler relationship. He had experienced this handover more times than I had, so the reality of my leaving seemed to wash over him. "Well, I wouldn't leave you with someone I don't trust," I said. He nodded slowly, seeming to appreciate my comments, regardless of whether they were truthful. And that was that. No more need to showcase the other guy.

"Unfortunately, you will have to continue working with this crook," I added, pointing to my linguist.

"And he's Egyptian," Jamal replied swiftly.

"Forget you both," said my linguist in English and then in Arabic. We all laughed.

I jumped back into a serious tone: "Thank you for your help. Thank you for trusting me with such important information." Jamal never broke eye contact. "I want this country to be free, for my kids and my grandkids. I've lived a hard life. I don't want that for them." He said all this as a tear rolled down one cheek. He hadn't intended that; it was as if the tear had snuck out. He quickly wiped it away. Those words might have come from my own father. I tried to remain stoic. "Well, we will continue to help you do that as best we can." He nodded graciously.

"What will you do when you get home?" he asked me. I pondered home for a second, long enough to have earned the attention of the whole table. "Probably eat some Cuban food and play some baseball," I said. He smiled inquisitively, as if interested in hearing about these two things wholly unfamiliar to him. It seemed he would ask more questions, but he held back. I interrupted the silence by saying that we were also traveling through Germany on the way home. This caused him to light up.

The 2006 World Cup had begun its final games in Germany, where the championship match would be played. Many nights were spent out on Bucca Square those final weeks, watching the famed international competition. The locals had the feed beamed through a satellite and onto a projector, which spit the live images onto the broad side of a cement T-wall, the same kind that surrounded the square. One of them brought in stereo speakers and, with that, we had a setup that would rival a theater in the States. The atmosphere was intoxicating. Cheering regularly erupted with such volume that it could be heard across the base. Very few Americans were there, so I had the fascinating privilege of watching countries compete among a foreign audience. I learned then that these Iraqis didn't favor the underdog, as Americans do. When I cheered for an upset, they virtually scolded me. They cheered for the team that was more likely to win.

Jamal had a twinkle in his eye. "My heart is in Germany," he said in nearly flawless English. I looked at him bewildered. "You said that perfectly, like an American!" My linguist, equally shocked at Jamal's command of English, translated quickly. The elder man waved us both off and replied, in English, "That's all I know." And he quickly looked away, smiling to himself.

We spent the next few minutes discussing soccer and the potential for Germany to win the cup in Germany, since they were one of only four teams remaining. What a finish that would be, he commented. He shared stories of his grandson, who was now kicking a ball, and his hope that one day he could play on the world stage. We spoke briefly of politics and of family; I had long ago fabricated a wife, to help me look more credible.

After about twenty minutes, I stood up and stuck out my hand. Jamal shook it kindly but firmly, as he rose slowly from the bench. We held that handshake for a minute, adding a few slaps on the back. "Sorry to bring you out in the heat of the day. It's just safer this way," I said. Jamal waved me off. "This is my country. This doesn't bother me." Good, I thought to myself, because I will take the South Dakota winters over this heat anytime.

Similar exchanges took place with other sources. Each required a different procedure. Each was emotionally draining in its own way. But the exchange with Jamal was the only one that I remember as a moment of growth for me as an intelligence officer and as a man. Lord only knows what has become of Jamal or of his grandson.

It was now three days and a wake-up, as we say in the military, until Erin and I would board the bus for Kuwait. For the first time in weeks, I slept soundly.

CHAPTER 22

The Finger

We loaded into the rhino, the term for the up-armored bus that transported personnel through some of the most inhospitable areas of southern Iraq. The thick plated armor that adorned the sides and bottom of the bus were put in place to theoretically withstand an exploding IED, and were resistant to gunfire. The buses were equipped with reinforced steel to absorb the blast and shrapnel from the deadly and frequently used roadside bombs. It was an ugly, slow-moving piece of engineering, but I cared only about its effectiveness. Erin and I had a drive of thirty minutes or so out of Bucca and over the border into the safety of Kuwait. But we had to go through Safwan, one of the deadliest cities in southern Iraq.

We heaved our bags into the lower compartments situated on the bottom frame of the bus, between the wheel wells. The day was incredibly hot. Having already moved eight bags from the rear of the base, I was sweating uncontrollably. But my mind was fixed on one last short trip to safety.

I assumed this trip would not go as smoothly as advertised. I stumbled to the back of the bus and plopped down on the second to last row. I took the window seat, gently leaning my head against the glass, feigning exhaustion in order to conceal the excitement welling up inside. The driver had the AC on high. The blast of cold air felt wonderful, after hauling bags across the base and languishing in the hot sun for hours waiting for our ride. I recalled then how the sweat-drenched long-sleeve shirt and gloves I had been required to wear as a goalie on my high school soccer team had kept me cool during those hot Florida afternoons. The soaked outerwear felt great on my hot upper body when the breeze blew, and the extra clothing shielded my skin from the sun.

Erin sat down beside me, equally thrilled but also doing her best to restrain her excitement. "Well," I said, without making eye contact, "Let's hope we don't get blown up on the way out." I could feel her staring daggers at me.

"Why would you say that?" she demanded.

"Just to get under your skin." Finally looking over at her, I now saw the look on her face. Without saying another word, she looked away.

The rest of the soldiers filed into the rhino, and the driver shut the door behind him. Silently and gingerly, he took a seat in the chair and began punching buttons on the dash. The ride was about to begin. Either we were going to make it over the border without harm, or not. We were helpless. Engaging the enemy would be nearly impossible unless we could all file off and out, into the line of fire. We had convoy security to the front and to the rear. But we were tucked rows deep, inside an up-armored bus.

I began to feel extremely uncomfortable, but there was nothing I could do. I had to concede control of the situation to the security around us. I had to trust they would do their jobs. Having traveled this unfavorable stretch of road before, I also knew that a U.S. Army checkpoint, Camp Navistar, sat just over the border in Kuwait. Between Bucca and Navistar, we had quick reaction forces that could respond from both directions. I took solace in the knowledge that we had backup, if it came to that.

Now my mind raced. I replayed different scenarios in my head. My training kicked in, as I evaluated the circumstances. I considered egress options, should an IED strike the vehicle. I examined the windows for ways to dislodge the glass and engage in a firefight. We wouldn't want to all pile out the side door. That would be like shooting fish in a barrel. Two airmen would be seriously injured two months later by an IED explosion that rocked Safwan.

The engine roaring to life shook free my mind. I just pushed the scenarios out of my head and tried to relax.

A low rumble shook the vehicle. The "brick," or walkie-talkie, resting on a cup holder near the steering wheel, chirped to life. "Packs loaded?" came a voice from over the brick. The driver casually grabbed it and responded with little enthusiasm. He clearly had practice in this transport exercise. It was often strange to see the lack of excitement from those who routinely participated in the warzone processes. It seems like everything should come with a certain level of drama. He seemed apathetic, but nothing could be further from the truth. "All accounted for, ready to depart," he replied in a monotone. A few more garbled transmissions came across, and we were off.

Each second felt like an eternity. Then again, everything happened quickly. Time stood still, even as it flew by.

The bus made its way along the serpentine route leading to the front gate.

Through the dusty, inch-thick glass of the windshield, I made out the guards up ahead. As we approached, I saw them waving us through. I could also see the long line of semis and other vehicles waiting for approval to enter Camp Bucca. The thought that I no longer had to concern myself with the security threats posed by visitors and vehicles, or any other threat for that matter, gave me rest. It also made me feel guilty. Counterintelligence responsibilities for the base now resided with others, but for some reason my departure made me feel I was somehow shirking my duties. The base would continue without me, but leaving now felt like I had not completed the mission.

A mental note from a past instructor came rushing forward: "Remember, you contributed to the war effort, but you will not win the war, and you probably won't be there when it ends. The intelligence cycle will keep going. Sometimes you will see the fruits of your labor. But often you will not. You must be able to pass it on to the next guy, sometimes without ever knowing the outcome."

Instinctively, as quickly as I could, I scanned the trucks. Perhaps

I could stop one more threat. Fight these guys to the bitter end. But my eyes glazed over. It was done. The gate arm rose just as the bus approached and, before I could blink, we took a sharp right turn and found ourselves on the main highway headed for Safwan.

The bus ran smoothly, despite the bad roads and the thousands of pounds of armor that covered its exterior. The ride was surprisingly comfortable. Whoever made the shock absorbers for this vehicle deserved a medal. The behemoth just pushed forward as I peered out the small window. We approached the overpass, the first dangerous point after leaving Bucca. IEDs could be concealed well, among the layers of bridges and the surrounding brush. And there was simply not enough time for a driver to recognize a threat and adjust at this intersection. Advance teams often disallowed convoys passage until they could deem the area clear.

Thankfully, the convoy traversed the overpass without incident. And now the bus lumbered toward Safwan. As we approached, I could see the mud huts dotting the sides of the road. Our arrival in town was preceded by Humvees taking up strategic locations on the roadside. Those manning the gun turrets turned their backs to the convoy, aiming their machine guns in the direction of the houses about fifty yards off the road. We flew by as airmen scanned the horizon, ready to engage any threat that popped up. Meanwhile, sirens warned drivers to pull to the shoulder or be squashed.

Homes and buildings increased in number the closer we got to the city center. There were now structures everywhere, mud huts gradually replaced by cramped cement structures. Some walls had bright colors with advertisements. Poorly drawn pictures of leaders and the Iraqi flag were sloppily applied to the sides of modest buildings. We passed what looked like a rudimentary auto repair shop, a bakery, and what could have been mistaken for a billiard hall. I tried to take it all in as we zipped down the corridor that led directly into Kuwait.

My ears caught the furious clip of a camera, and I turned to see a female airman in the rear of the bus snapping photos as if her life depended on it. I hadn't considered photography, despite having my camera in the front pocket of my vest. Photography sometimes felt to me like it distracted from the moment. Throughout my deployment, I favored letting myself become absorbed in the colors, the textures, the smells of the moment, then relying on memory. That airman told me later that she had never left Bucca – never had a reason to, because of her job.

Another Humvee parked on the side of the road was in my sight line, then just as quickly out of it. The insulated bus prevented me from feeling the sun's heat. It hadn't dawned on me that a Humvee heightened my senses when driving through town. The heat, the stench of gasoline mixing with food and other smells wafting through town, all of it was unavoidable in Humvees and other military vehicles, and strangely enjoyable. The interior of the rhino offered no such exposure. Outdoor odors were replaced by the scent of cleaning solutions and recycled air.

A few more minutes of driving, and we would be out of the danger zone. It was a short distance from Bucca to Navistar. A loud crunch erupted nearby. And the bus continued forward as if nothing had happened. If there was any extra activity in front of or behind us, I couldn't tell. The roads were so poorly maintained that it was hard to tell. I chalked it up to running over one of the massive pieces of debris that commonly line the roads in Iraq.

A few minutes later, we crossed the border into Kuwait and took a left into Navistar. Our bus came to a complete stop in a long line of other vehicles. I assumed that, since everything in the military is hurry up and wait, we would be waiting for hours to exit the bus. But just as I thought it, the driver stood up and pushed the doors wide. "Welcome to Navistar. Come unload your gear, and make sure you check in with your command." And then it sank in: We had made it. I could actually breathe a sigh of relief. For the first time in months, I let my guard down and allowed myself to relax. Erin and I looked at each other and smiled. No words were needed. And we certainly weren't completely in the clear. Yet this was a big step toward returning to normal life.

We exited the bus, gathered our bags, and made haste to the pickup location where our officemates from Kuwait awaited us. I could see in the distance the two black Suburbans. My boss, his assistant, and other agents were lingering next to the vehicles as we approached. Ramel, my tall Sudanese linguist, was the first to greet us. "David, look at you!" he bellowed in a thick accent. "Beard and weathered skin – you look Syrian!" Other agents greeted us warmly as we threw our gear into the SUVs.

I hadn't noticed the disappearance of the boss until he came running over from having spoken with the convoy commander. He hurried over and exclaimed, "Are you guys all right?"

I looked at Erin and back at the boss, confused. "Uh, yeah, we're fine."

He paused. "Wow, okay. Because I just got word that the bus got hit by a Coke can IED."

Terrorists had been using soda cans filled with explosives to hurt soldiers patrolling the roads on foot. Easy to make and easy to transport. The small-scale bomb proved more a nuisance than anything else. But it could do some damage to unsuspecting security walking the streets.

That revelation caused me to snatch the shades from my face. I looked at Erin and back at the boss. "Are you kidding me? I thought I felt …" Erin jumped in, explaining how we thought we had felt and heard something. We agreed among ourselves that the sound we had heard had probably been that tiny, insufficient weapon.

And then I let out an enormous laugh. "Unbelievable." I shook my head and looked back in the direction of the border, my eyes trained on the distant landscape. Iraq had just given us the finger on the way out.

About the Author

David A. Grantham serves as a Senior Fellow with the Center for a Secure Free Society, where he researches and publishes on regional security matters related to the Western Hemisphere. Prior to that, David served as Senior Fellow of National Security at the National Center for Policy Analysis (NCPA) and as an officer in the United States Air Force with the Air Force Office of Special Investigations (AFOSI).

David holds a Ph.D. in History from Texas Christian University and a Master of Science in International Relations from Troy University. He earned his Bachelor of Arts in History from the University of South Florida in 2004.

He spends his free time exercising, reading, and experimenting with all types of new foods. David resides in North Texas with his wife and children.

His work can be found at granthamstrategies.com

About Blue Ear Books

In 2009 journalist Ethan Casey, author of the landmark travel book *Alive and Well in Pakistan* (praised as "magnificent" by Ahmed Rashid, author of *Taliban* and *Descent into Chaos*), decided to take control of the publication and promotion of his own books, rather than remain subject to the whims and vicissitudes of a perpetually and radically changing publishing industry.

That led to the founding of Blue Ear Books, through which Casey and a small group of colleagues publish a select list of books by authors with distinctive professional and personal perspectives on the international world in the 21st century. To learn more, visit:

www.blueearbooks.com